Pride of the line – No.71000 Duke of Gloucester *heads the 13.50 Rothley – Loughborough out of Quorn & Woodhouse on 23rd November 1986.* *(J. B. Gosling/GCR)*

Echoes of the GREAT CENTRAL

John M C Healy

Haynes

GREENWICH EDITIONS

GREAT CENTRAL
This edition published 1996 by
Greenwich Editions
Unit 7, 202-208 New North Road
London N1 7BJ

First published 1987

ISBN 0-86288-076-9

Printed and bound in Great Britain

Contents

Introduction

Of all the major new main line routes built in the nineteenth century, the Great Central seems to attract the most attention, since its London Extension line was the last to be built. This was in an age when the motor car was already beginning to appear. In fact it was to be the motor car that eventually put pay to the majority of the Great Central's traffic as competition from road traffic increased during the last 15 years of the line's life. The Great Central was so named because its London Extension line cut through the very centre of the English countryside, in an area hitherto untapped by the other main railway companies, who provided stiff opposition for the scheme which was the brainchild of a Victorian entrepeneur, Sir Edward Watkin.

Probably the most interesting part of Sir Edward Watkin's ambitious scheme was his desire to link major trading centres in England with the Continent, via a Channel Tunnel. In light of recent parallel ideas to build a link between Britain and the Continent it poses the question as to whether the Great Central London Extension would still be in existence today had the Channel Tunnel been completed earlier this century. It is probably true to say that the tunnel would have been the key to the line's survival in the sixties when the 'Beeching Plan' removed the line from the network. The reasons behind the closure which will be covered more fully in the first section of this book, were mainly due to the fact that the Great Central was a duplicate main line which carried very few passengers and was considered of little value commercially because of the areas it passed through.

The aim of this book is to provide an account of the line under construction, with a number of illustrations taken from the Newton Collection at Leicester which depict various features and aspects of the line during the period 1894 to 1899. Then to follow the story of the line, choosing various topics from stations and their staff, locomotives and the trains they operated plus a final trip down the line to the eventual closure and what effect it had on the towns on the route. Unlike the many fine works on the Great Central which have already appeared portraying it as it was in its heyday, I have endeavoured to include unpublished material on the London Extension and to provide a detailed comparison between the original Great Central and the Great Central Railway of today. The latter came about as the dream of a few local inhabitants of Leicester, who like Sir Edward Watkin, had the desire to own and run their own stretch of the Great Central through the heart of the Leicestershire countryside. The dream like Watkin's was ambitious and costly, and in the end the members of what was called the Main Line Preservation Group had to settle for the lesser option open to them, which meant the preservation of only one of the two tracks between Loughborough Central and Quorn & Woodhouse. This was later extended to Rothley, but sadly neither the section between Rothley and Leicester, nor that between Loughborough Central and Nottingham could be retained at that time.

The Great Central of today boasts one of the finest collections of steam and diesel locomotives, a good catering service and stations with all the old traditional features of a byegone railway. It is therefore true to say that the ways of the Main Line Steam Trust are all 'Echoes of the Great Central'. In choosing this title for the book there were two significant factors which I felt made it appropiate in as much as the Loughborough organisation is an echo of the line's former past. The second reason is that someone once said to me that even though the railway had left their town, it was still possible to hear the steam engines at night...

Ex LNER N2 class 0–6–2T, No.4744 arrives back at Loughborough on a sunny afternoon, 3rd May 1980. Under the restored station canopy the buildings are, left to right: the gentlemens', a store, the waiting room, the second hand shop, refreshment room, stationmaster's office and the souvenir shop.
(P. D. Nicholson)

Prologue

We shall start by taking what is described by the present day Great Central Railway as a Return Trip to Yesterday. This is a nostalgic journey from Loughborough Central station on the old Great Central London Extension, for a distance of about 5 miles southwards to Rothley, and return.

One is first aware of the Great Central's presence when arriving at Loughborough Midland Station which is a rather stark structure dwarfed by the overpowering shadows of the Hawker Siddeley/Brush Engineering works, famous for the building of locomotives. Skirting round the face of the Brush buildings are the sad vestiges of a once great main line which is now reduced to the status of nothing more than a long siding for British Gypsum trains. This siding is connected by a spur to the Midland line south of British Rail's station, at which point there used to be a bridge that echoed to the rumblings of 9Fs on 'Windcutters' or B1s on semi-fasts or 'Directors' on Manchester to Marylebone expresses. Gone sadly are those days-or maybe they are not after all, because if the Great Central Railway Company (1976) succeeds in rebuilding the bridge over the Midland and extending to Ruddington, then it will be possible to see all the above engines working trains once again over this spot.

For the moment Ruddington is only a dream, like the London Extension was ninety odd years ago in the eyes of Sir Edward Watkin, its founder. Fortunately some steam addicts in an unparalleled effort have saved a part of Sir Edward's dream and this can be found on Saturdays, Sundays and Wednesdays (during Summer) by the tell tale shrill whistle and heavy palls of smoke which pour forth into the air, which always act as a good pointer when wanting to find Loughborough Central. As one leaves the main A6 road walking towards Great Central Road there is a strange sensation of stepping back in time as one passes row upon row of terraced workers' houses with the Great Central Hotel sitting snuggly amongst them. Opposite the hotel stands the old station goods office and the site of the yards which have been swallowed by industry. Looking up the hill, as if proudly proclaiming 'this is the Great Central' stands the entrance to the station and booking office. Although the bricks are blackened by generations of steam smoke, the date of its construction is still visible – 1898, one year before what was destined to be the last main line in England was opened.

Entering the booking hall and going to the ticket office there is a strong temptation to ask for a single to Marylebone or a return to Nottingham Victoria, as it is hard to imagine from the way the upper part of the station has been restored to its original teak, that the station has ever been closed for

Right: Still retaining much of its original character and style is the Great Central Hotel as seen in May 1986. Once the Hotel provided a welcome resting place for weary travellers though now it serves as a popular hostelry for volunteers running the preserved section of the line between Loughborough and Rothley.

(Author)

Right: The booking office and Manager's office at Loughborough as seen from Great Central Road on a summer's evening in 1986. The interior of the booking office has been lovingly restored with its original wood panelling, while the canopy has since been attended to.

(Author)

Far right: The ticket window at Rothley, which is typical of all those to be seen on island platform stations, the length of the London Extension-complete with clock, gas lamp and billboards.

(Author)

A RETURN TRIP TO YESTERDAY
ON SALE HERE

GREAT CENTRAL RAILWAY

The well kept station buildings, platform and canopy as seen from the trackside by platform 1 in late August 1986. The hanging baskets, seats, enamel signs and other fittings conjure up scenes of the past and the train could be waiting for a change of engine before working to Marylebone!
(Author)

long distance travel. Sadly there is no chance of seeing the fine scenery which the railway passed through in Nottinghamshire or Northamptonshire. Still, a trip to Rothley and back will no doubt be sufficient to capture some of the magic atmosphere created by the steam train and the old Great Central. After purchasing a ticket and walking down the main staircase to the platform, one is immediately aware of one of the Great Central's hallmarks of construction in that the platform stands in the middle of the main running lines. Also apparent is the fact that Loughborough must have been regarded as an important station, hence the number of buildings under its massive canopy which has been painstakingly restored to its original condition.

In front of the staircase a large platform sign with white letters on green backing displays the name 'Loughborough Central', while underneath there is a billboard with various timetables and posters recreating the scene as it would have been in GCR and BR days. Two of the timetables show 1957 and 1958 departures for Nottingham Victoria and Loughborough Central respectively. Copies of these timetables are on sale in the shop which is an Alladin's cave as far as railways are concerned, with a full range of videos, tapes, books, cassettes, badges, ties, postcards, posters and miscellania. Once a waiting room, this room now has more interest taken in it than ever before and one cannot help but wondering whether the station at Loughborough would not have benefited from a shop in days gone by. On this one can only surmise as one strolls down the platform, past the station master's office to the refreshments room which has had the same role for nearly 90 years. Here, it is possible to purchase anything from a sandwich to a cooked meal and this is one of Loughborough's most needed facilities-catering for the hungry visitor or the ravenous volunteer.

Moving down the station platform to the second block of buildings one cannot help but notice the rake of coaches in the sidings, either side of the main running lines in 'preservation row'. Aside from the state of these carriages they could quite easily be stored, waiting to form an excursion to Mablethorpe, Sutton-on-Sea, Bournemouth or Swansea. Alas, this motley set of Gresleys, BR Mark 1s and other assorted types will not see service for a long time to come. This will probably be when the Ruddington Extension is completed and there are proper facilities for the Carriage and Wagon Department who are compelled to fight the elements in order to effect repairs.

The second block of buildings on the platform is interesting in the sense

The most impressive part of Loughborough Central's platform structures is the canopy which has been lovingly restored.
(Author)

that one part of this has been restored to its original condition with teak varnished panels and seating. The Victorian atmosphere in the ladies/waiting room is further enhanced by the picture of Queen Victoria and the picture of Sir Sam Fay, the Great Central's General Manager. To complete the facility, on the table in the centre of the room, magazines are laid out for people to read. The latter room was in two parts; a waiting area for the third class, the rest being the first class ladies room. The first class section is now used for another enterprising venture on the Great Central in the guise of a second-hand stall, which regularly earns the company a revenue of £2–3,000 annually. At present, the end part of the second block of buildings is in a poor state of repair and is used as a store room for the station restoration team. In time it will be restored as a further waiting room. The last building on the platform, facing north resembles a shrine in construction as it stands on its own away from the main blocks. This is the gents' toilet block and retains most of its original tiling and timbers, though the urinals themselves have long since had modern fittings installed!

The next port of call is the locomotive shed which is reached by a short walk down the path alongside the carriage and wagon siding and shunting loop. Going along the path the goods shed is visible, and which now belongs to 3M Health Care and is used as a store, while the old goods yard has been turned into a car park. The entrance to the yard can still be determined by the water tower which stands further down the track. The entrance to the yard is also traceable by the way the embankment curves round into what was until recently a fenced off enclôsure for Leicester Heavy Haulage. The proximity of their yard was a bonus for the Great Central, especially when the delivery of new engines took place. Mostly though, additional locomotives are taken by road to Quorn unless transportation difficulties are experienced, as was the case with No.40 106 *Atlantic Conveyor*. In front of the yard the turntable obtained from Calais lies in pieces, alongside the remains of one of the Main Line Steam Trust's engines, *Hilda*. Also lying in store are a number of signals acquired from Toton for the equipping and controlling the Quorn Loop.

Above left: The goods offices of the Great Central, and stable which was to the left of this building, now serve as offices for 3M Health Care who also own the old goods shed at Loughborough. The gates on the right form the entrance to the old goods yard while the original weighbridge and hut still survive.

(Author)

Above right: The restored signal cabin which has been transformed from an empty shell to a fully commissioned signal box. This was brought about by the installation of equipment retrieved from Ruddington.

(Author)

Approaching the shed, the smell and taste of steam crawls down one's lungs and has the mesmeric effect of some addictive potion. But before being lured away by the smells of simmering steam engines there is just time for a visit to the signal cabin which is the nerve centre of operations, controlling all the points, signals and ground discs in the station and yard area. The signal cabin is one of the most fascinating areas of the preserved line, with all its shiny polished levers, clocks and numerous block instruments. To break the silence in the signalman's lonely job the cabin echoes out with a number of bell rings indicating the approach of a train into the station. As the signalman sets the points for the train to draw into platform 2, a couple of extra levers are drawn forward to give the train a clear path.

Having watched the train successfully signalled into the station it is time to continue on down to the shed, where a trip round to the back is always first on the agenda. At the most northerly point, the past and future are evident as the old trackbed abruptly ends with a bridge that goes nowhere. The reason for this is that a large section of the old embankment has been removed for other uses. In the distance is the formation of the line to Ruddington now connected to the Midland main line to London. All being well the Great Central will be running to Ruddington by the year 2000, once the bridges and embankments have been reinstated. One can only dream what it will be like to see Great Central engines pounding over this section again, which judging by the success of the recent promotion, will see work started before too long.

Behind the shed it is a maze of van bodies containing parts, spares and tools for the engines which are in pieces forming the overflow of restoration projects. These locomotives are the 9F which is kept warm and cosy in a long tall single span shed; the Thompson B1 No.61264 being treated under canvas and No.48305, which rather amusingly carries a sad looking face on it with the words 'Please don't let me die' painted on the smokebox door. In addition to the steam engines a couple of diesels, namely *QWAG* and No. D4279 *Arthur Wright* sit, inactively on a siding.

Moving inside the shed, amidst the tools, lathes, forges etc, there is a

Looking to the future, the railway will eventually extend northwards from Loughborough Central to Ruddington. North of the bridge in the foreground, the embankment will need replacing and a bridge over the British Rail main line will need to be installed before the extension can be completed.
(Author)

The boiler of LMSR Class 8F 2–8–0 No.48305 after removal from the main frames which took place on 14th April 1986. Fortunately the witty sentiment on the smokebox was granted and the locomotive now has a secure future.
(Author)

selection of steam and diesel locomotives large and small. Two of the occupants of the shed are *Boscastle*, which lies in an unfinished condition, mainly because its owners have not yet raised all the necessary funds to enable completion, the other being No.5231 whose boiler is still in the process of being worked on prior to assembly on to the main chassis. Apart from these locomotives are the recently restored *Duke of Gloucester* and *Witherslack Hall*. The shed is also used for storing locomotives when they are not in service and on a typical visit can be found No.506 *Butler Henderson*, Y7 No.68088,

No.1306 **Mayflower** *outside the shed at Loughborough awaiting its next turn of duty, 3rd May 1980.* *(P. D. Nicholson)*

Robert Nelson No. 4, *Hunslet 1800 of 1936 from NCB Littleton Colliery, Cannock, Staffs by the Loughborough engine shed, with the Thos Hill diesel on 4th July 1981.* *(P. D. Nicholson)*

GWR No.5224, diesels Nos14 029 and 40 106 and three engines being prepared for the day's services such as No.1306 *Mayflower*, Austerity No.68009 and N2 No.4744, with *Robert Nelson No.4* on standby.

Wandering out of this home of lovingly cared for giants it is almost impossible to believe that the original main line went straight through where the shed now stands. Once the extension to Ruddington is complete the shed will either be dismantled and re-sited, which will have to be done with the 9F shed whatever happens, or measures will have to be taken to slew the main line round the side of the shed and onto its natural course, which would cause less disruption all round.

Leaving the shed area there is a chance to have a final look at the line up of rolling stock before making one's way back along the path to catch the train to Rothley. Before boarding the train though, a visit to the Museum is a must. Once inside, the four walls of the old lift shaft shout out the story of the line's construction on picture boards which display photographs from the famous Newton Collection. In addition to these there are a huge number of rail chairs, the odd section of rail, mileposts and even a contractor's hand-shovel. All the Museum needs is an original mechanical excavator to complete the display, though space would prevent this! The second room makes one feel dwarfed by its height and is like something out of Alice in Wonderland with its impressive array of station signs, bridge plates, shed plates, posters and other railway paraphenalia. These items are laid out and fixed to the walls which represent the colours of the original Great Central livery, before the adoption of teak for its carriages. The display cabinets around the room house various smaller exhibits which are changed frequently to show the vast range of items in the care of the author, who is the Museum Curator. Most notable

The then newly refurbished Museum, September 1986, showing a signal level frame, locomotive shed code plates, carriage boards, posters, a level post and enamel signs. ***(D. Bonas)***

of all is the collection of plates, cutlery, tableware, badges and paybooks recording workers' wages. Two of the cabinets however contain special exhibits, one holding an excellent working model of a 4–4–2 Atlantic, No.361 resplendent in Great Central green livery, while the second case contains an 00 gauge scale model of Quorn & Woodhouse. This is in the process of completion and will eventually resemble the scene at Quorn on Road Rail Steam days.

When visiting the Great Central it is always wise to visit the Museum prior to boarding the train, as one can time one's stay by listening out for and watching through the windows for the locomotive backing on to the train. This means there is just time to hop on to the train prior to departure. Having a last look at the platform from an open window one would think the station would be a good example of a title for an H. G. Wells style book, 'Time has Stood Still', judging by the beautiful hanging baskets suspended from the canopy and the antique enamel advertisement signs that proclaim long extinct products, as well as the staff who are all attired in a period uniform.

Suddenly, the train lurches forward on the shrill blast of the guard's whistle and we're off to take a train ride through the beautiful Leicestershire countryside, hauled by *Mayflower*, one of the Railway's splendid steam locomotives. As the train eases out of the platform and under the bridge, the carriage becomes filled with smoke, and it is quite easy to believe that one is

***The main bridge at Loughborough Central which was partly demolished by an errant road vehicle and subsequently repaired, hence the different colour brickwork.* (Author)**

aboard an express passenger train for London Marylebone back in the 1950s when the Great Central was still open as a main line from Manchester. Originally trains attained speeds of up to 80mph due to the care taken during the line's construction to ensure that no gradient was greater than 1 in 176 and that there were no obstacles allowed to get in its way, such as level crossings. On one side the loop branches off to platform 2 and once the carriages have clattered over the points, another famous landmark looms up on the left in the form of Ladybird Books. This has been aiding childrens'

The line of coaches in the 'up' siding awaiting restoration. From left to right these are LMSR all steel brake third (Built 1949), GCR 'Barnum' open saloon No.5666 and LNER corridor brake third No.16520. On the right is the run-round loop and associated signalling, that on the left being the 'up' starter. **Duke of Gloucester** *is seen passing the Ladybird Book premises having just set off for Rothley on a trial run.*
(Author)

learning, with its comprehensive series of educational and story books which have been produced since the First World War. The Company now produces over 20 million copies each year in over 60 languages even though they only possess a staff of 285.

Once past Ladybird's factory the train steams through a residential area with a new housing estate on the left, before reaching the bridge that leads to the Charnwood Water beauty spot and park. Over the years various decisions have been made concerning the erection of a halt at this point although there are two disadvantages, namely that it might preclude the laying of a second track and also the site is too close to Loughborough Central. The next bridge passed is the large steel girder bridge carrying the A6 road over the Railway, and now one can enjoy a leisurely Return Trip to Yesterday, passing through the Charnwood hunting country. Woodthorpe comes into view soon on the right with the bridge leading to the village itself. From here, the line continues on a rising gradient with rolling hills on either side and Charnwood Forest in the distance on the right. Straight after this a small stream is crossed on a high embankment, then the tennis courts of the Manor House Hotel are seen on the left and the train drifts into Quorn & Woodhouse station, which is a perfect spot to disembark from the train and have a picnic or maybe explore the famous Quorn hunting country.

GENERAL
WAITING ROOM
WAY OUT

Left, top: The approach to Quorn & Woodhouse with the new loop in situ. Before this became operational the track had to be ballasted and signalling installed.
(D. Bonas)

Left, bottom: The road overbridge at Quorn & Woodhouse was built by the girder method rather than the brick arch method as at Rothley. The picture also shows the way that the staircase was built into the bridge.
(D. Bonas)

Above, left: The 'up' platform at Quorn & Woodhouse. The building furthest from the camera is the gentlemen's while the building nearest the camera contains the shop and general waiting room.
(Author)

Above, right: Swithland, once an important exchange point with the Mountsorrel quarries and a proposed site for a station. On the right is the course of the old 'down' line and in the centre is the intended access to the station site. Plans for the future include the building of a station here and the transfer of the Road Rail Steam event to the old yard.
(D. Bonas)

Leaving the beautifully restored island platform at Quorn & Woodhouse, the train passes a row of goods wagons stored in the siding at the end of which stands the Great Central's latest engine acquisition, No.35025 *Brocklebank Line*. This is in a very poor state indeed and will require a tremendous amount of dedicated restoration. Indeed the same effort will be required on the signal box which lies at the end of the station yard. On the opposite side, neatly arranged in the bank are a series of whiter than white letters showing the name Quorn and Woodhouse. Once over the newly installed point for the loop we are finally bound for Rothley the present terminus of the line. Having cleared Quorn the train proceeds past Buddon Wood on the left while the Mountsorrel granite quarries lie further beyond. On the other side there are further panoramic views of the Charnwood Forest area with Beacon Hill, some 248 metres above sea level, on the horizon. Steaming along the line the train continues on a steadily climbing gradient on what is known as the Quorn Straight, surrounded by fields and pastures that resemble a patchwork quilt with their different colours each separated by long green hedgerows. Suddenly, after a series of wooded copses and fields the line swings gently to the left, passing under the bridge carrying the road to Woodhouse Eaves. The line then hits Swithland Reservoir which lies either side of the two massive viaducts that carry the railway across it. The reservoir is surrounded by trees and the views from the train are some of the best to be afforded of the whole area. As the train steams across the first viaduct we arrive at Brazil Island before being borne over the second viaduct. One advantage of today's ride is that the train allows one to have a better look at the surroundings due to its rather leisurely progress, and of course there is always a chance to see the views missed on the way out when the train returns from Rothley. During the construction of the railway the reservoir, which is responsible for supplying Leicester with water, had to be emptied to allow the viaducts to be erected. Because of the vast quantity of water and the surrounding woodland, the area is an ideal habitat for a wide variety of bird life.

Leaving the reservoir the line crosses the Swithland to Mountsorrel road at which point it is evident from looking down the gap between the two spans that carried the 'up' and 'down' main line, that something exciting once happened here. There is a bricked up archway and stairs built into the bridge that lead up to the trackbed. Apart from this the parting of the lines show that there was once space for a station here, that was proposed but never built. However, this station area may yet be developed to fulfil its original role, providing an alighting and boarding point for people wishing to take full advantage of the scenic delights. Swithland, as can be seen from the vast open space opposite 'the station' area was once the site of a number of exchange sidings and a signal box for the handling of traffic from the Mountsorrel Granite Quarry Railway whose overgrown trackbed curves off to the left.

The main line curves to the right and the approach to Rothley is evident as a number of houses come into view on the left. After this we begin a gradual descent to the actual station which is the terminus of today's operations. As the train draws to a halt Rothley comes alive with activity as passengers get off in order to have a better look round this quaint little station. With ten minutes to spare there is just time to have a quick look at progress on the Birstall Extension. So far there are huge stocks of materials for this major

Below, left: Rothley showing the formation of the permanent way under the road over bridge and round the platform. ***(D. Bonas)***

Bottom, left: The gentlemen's at Rothley was typical of the Great Central platform buildings for country stations. These were designed to be built with the minimum amount of materials. ***(Author)***

Bottom, right: The refreshment Waiting Rooms at Rothley are nearest the camera, while the ticket office and stairway to the road are in the distance, right. ***(Author)***

Right, top: A porter's barrow at Rothley, complete with suitcases and trunks. ***(D. Bonas)***

Far right: The end of the line at Rothley – but not for long as efforts are being made to extend the line towards Leicester as far as Belgrave & Birstall. ***(Author)***

feat which will eventually provide the Great Central with an extra $2\frac{1}{2}$ miles of operating line. In preparation for this a signal box rescued from GC territory has been restored, rebuilt and painted dark green with white window frames, making it look aesthetically pleasing. From Rothley the trackbed to Birstall passes over Rothley Brook and the formation becomes more sinuous and is overlooked by Thurcaston village which can be seen on the right. A mile and

***Below:* The signalbox at Rothley, which was recovered from Blind Lane on the GW/GC Joint Line is seen at an advanced stage of restoration in October 1985.**
(D. Bonas)

a half out of Rothley the line disappears into a deep cutting and a number of houses on the left are visible. Then, after passing under a farmer's bridge the formation curves right and drops down to Belgrave & Birstall station site on the very outskirts of Leicester. Suddenly, while I have been imagining what the journey to Birstall will be like, a large crowd has gathered to watch *Mayflower* approach the platform on the other side as it runs round its train. Once past us the engine is hounded by the camera clicking section of the group as if it were a celebrity or a member of the Royal Family.

Soon the train is ready to return to Loughborough and one has the feeling that the process is like winding back a film, as all the gorgeous scenery is seen in reverse order and even the engine is back to front until we reach our destination due to the lack of turning facilities. Sadly the end of the journey is soon upon us. Once off the train and we begin to make our way home, thoughts of the Railway's past are inevitably uppermost in one's mind. How did the railway come about? When was it opened originally? Perhaps most interestingly of all, why was such an apparently important line closed? Then, how has this part of the line been brought back to life? These are just some of the questions that will now be answered.

No.506 **Butler Henderson** *is seen in this September 1985 photograph running round its train at Rothley. In the distance are materials for the extension to Belgrave & Birstall.*
(Author)

Part 1
The Great Central London Extension 1893 – 1969

GCR No.1114 at Leicester on an Army special originating from the South Eastern & Chatham Railway.

Chapter 1
Setting the Scene

All photographs in this chapter are courtesy of Great Central Museum, Loughborough.

Like so many other railways built in the nineteenth century the Great Central owed its prosperity to one man, namely Edward Watkin. He turned an ordinary railway company running between Manchester, Sheffield and Lincoln into part of a larger empire called the Great Central Railway which he did by means of driving a new line from Sheffield through Nottinghamshire, Leicestershire, Northamptonshire and Buckinghamshire into London. This line was only a mere part of Sir Edward's plan as he intended to be able to run trains to the Continent by gaining a controlling stake in a number of railways which would give him a clear way to the Channel, under which he would build a tunnel. Besides this rather ambitious and indeed incredible scheme Watkin was responsible for the building of a structure similar to the Eiffel Tower at Wembley, although this was never completed, as well as two parks, one at Wembley and the other at Belle Vue in Manchester.

The sheer magnitude of his dream and the fact that the railway he constructed into London was the last main line to be built in England has resulted in Sir Edward Watkin attracting more of a 'following' than any other of his contemporaries save for Brunel. Sir Edward was born in Ravald Street, Salford, Lancashire in 1819, the son of a wealthy cotton merchant, Absolom who was noted for his involvement in the Anti-Corn Law League. His two brothers John and Alfred became respectively vicar of Stoxwold and Mayor of Manchester in 1873–4.

Returning to Edward, after a private education he entered his father's mill business and later, in addition to his business career he soon became interested in public movements and as director of the Manchester Athenaeum, he helped to arrange literary soirées. Like many contemporary prosperous Victorians of his social class, Watkin involved himself in many philanthropic schemes, among them the raising of money for the provision of Public Parks. As secretary of the fund raising committee he saw the opening of three parks in 1845, two of which were mentioned earlier. His literary and business interests led him also to found the 'Manchester Examiner' and by this time he had already become a partner in his father's mill.

At this point the start of Watkin's interest in railways showed itself when at the age of 26 he took on the secretaryship of the Trent Valley Railway which was later sold to the London & North Western Railway Company for £438,000. Soon after his involvement with the Trent Valley Railway Company, Watkin made a journey to the United States of America in 1851, an account of which he put into publication on his return in 1853. Following his return Watkin obtained his first major position of General Manager of the Manchester, Sheffield & Lincolnshire Railway under Huish, an old ally of his. This was a major turning point for Watkin as his career in the MSLR was to become a keystone in his great railway empire. At an early stage of his appointment he was already beginning to make suggestions that the Company ought to look at following the example of the other railways by extending to London. Had such a move been made at this time the Great Central London Extension and the Channel Tunnel scheme might have been more commercially viable.

In 1861, at the wish of the Duke of Newcastle, Watkin travelled to North America on Government business. This time he was sent to Canada to investigate means by which the then five British States could be combined to form a Dominion. He considered the first possibility of transferring the Hudson Bay Company to Government control, an expedient which was actually adopted in 1869. A further scheme was envisaged by Watkin by which Quebec would be connected with other Canadian ports on the Atlantic coast.

The years from 1862 following were to prove the busiest and most rewarding for Watkin as he obtained a number of new positions and titles in addition to those he already had. First in 1863 he gained a co-directorship of the Manchester, Sheffield & Lincolnshire Railway, then he was made Chairman the following year when he also received a knighthood for his services. By this time, because of his skill and enterprise in matters pertaining to railways, Watkin was already being viewed with the same esteem as George Hudson, the 'Railway King' who, two decades earlier, had more than 1,000 miles of railway under his control and had been elected MP for Sunderland. The Manchester, Sheffield & Lincolnshire line incorporated many smaller companies like the Cheshire Midland, the Manchester South Junction & Altrincham, The Marple, New Mills & Hayfield Junction and the Stockport & Woodley Junction.

In 1866 Watkin became a director of the Great Western Railway and in the following year gained a position with the Great Eastern Railway. He was also destined to gain the control of the South Eastern, Metropolitan and East London Railways. Thus Watkin by strategically 'acquiring' various companies began to lay the foundations for the fulfilment of his dreams – the construction of a railway from Manchester to Paris via a Channel Tunnel link.

In 1873 Watkin put forward a scheme which was a concerted effort to try to bring the Manchester, Sheffield & Lincolnshire Railway to London with the help of the Midland Railway. He proposed to construct a line from just north of Doncaster to meet the Midland at Kettering which, had it worked would have given the Midland direct access to the North Eastern and allowed the MSLR easy access to London St. Pancras. Alas the Bill for the scheme was thrown out of Parliament, leaving Watkin to look elsewhere to achieve his aims. Undeterred Watkin tried once more in 1877 to try to interest the Midland, only this time he involved the Great Northern also in the hope that the two companies would either be able to purchase or lease the Manchester, Sheffield and Lincolnshire. In the event it is likely that the sheer magnitude of his scheme was enough to ward off any existing railway company from becoming financially involved. Following this unsuccessful attempt Watkin tried to form an alliance with the Great Northern who had long since been an ally of his, but once again his terms were considered too high making the construction of a new line to London that much more of a certainty.

He began his task with the construction of the Manchester and Liverpool line in 1877 and the extension of the Metropolitan to Buckinghamshire which took place in 1892. Also as part of his growing empire, a line was built between North and South Wales, including the Mersey Tunnel which allowed a rail connection to be established with Lancashire. His schemes were not, however, restricted to British railways and one of the most interesting developments which he brought about abroad was the electrification of the Athens to Piraeus Railway in Greece. At this point it would be true to say that Watkin was at the prime of his life, but his crowning glory was still some years away – the passing of the Bill which would authorise the construction of the London Extension. On top of his business life he was a family man with a wife, Mary Briggs whom he married in 1845, and two children. Alfred Mellor Watkin became MP for Grimsby and a director of the Manchester, Sheffield & Lincolnshire Railway from 1875–1877 and 1899–1900, and daughter Harriette, who married Worsley Taylor of Moreton Hall, Whaley. Sir Edward's second wife, Anne Little died in 1896 three years after they were married. Watkin himself though did not live much longer as he died in 1901 before his ultimate ambition, the construction of a Channel Tunnel, could be fulfilled.

Watkin's predecessor on the MSLR, John Chapman whose protégé he had once been, was markedly different in character but was a keen supporter of his. Chapman was a patient negotiator and basically easy going in temperament, whereas Watkin was a commanding, ambitious entrepreneur. In politics they belonged to different parties. Chapman was Conservative MP for Grimsby from 1862–1865 and 1874–1877 until his death. Watkin, by contrast was a staunch Liberal MP for Great Yarmouth 1857–8, Stockport 1864–8 and Hythe 1874–95. Both men shared common interests in the literary and public world but their main interest was in the enhancement of the Manchester, Sheffield and Lincolnshire in which both men had large sums of money invested. If anything it can be said that the economic trends of the time were against Watkin and Chapman as they failed to increase the company dividend to shareholders and, because the time when the railways were prosperous and expanding had finished, they were unable to take advantage of the system whereby companies such as the Midland rented lines from each other.

Indeed by the time of the MSLR's proposal to extend to London was introduced, the boom period for the birth of new lines was almost certainly over, which made the latter company a far from prosperous concern, unlike the rival Great Northern and Midland companies. These two companies believed

The New Line to London

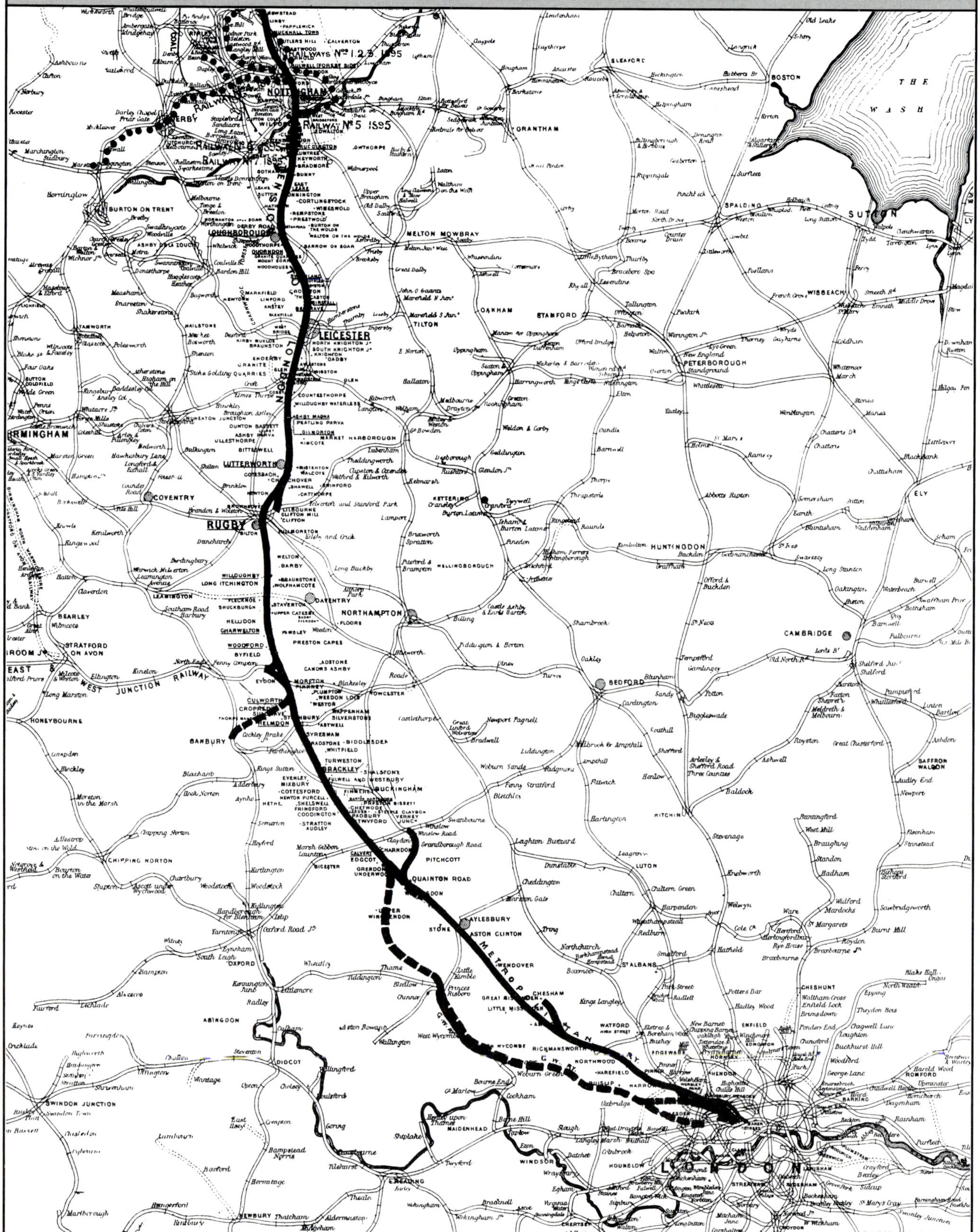

quite rightly that a third route to London would be completely superfluous and therefore uneconomic, since the major towns were already served by the existing network of railways. Thus the route for the Manchester, Sheffield & Lincolnshire was predetermined and lay through the heart of England, which, at that time, was mainly agricultural and served by a small number of villages. Watkins' logical argument for the construction of a new line was that there would be sufficient growth in traffic to justify such a route as he reckoned the other lines were by now worn out and in some respects he was right.

He further asserted that the other companies would need to expand and improve their lines. In this light Sir Edward set about his gargantuan task to build the new line, but he had failed however to take into consideration that the age of the 'horseless carriage' had arrived and to realise its likely effect on the long term development of the railways, even though the full impact was not felt until the construction of the motorways after the Second World War. Watkin also claimed that the Great Central would be a positive benefit to London since it would transport more supplies like milk, coal, timber and food generally from the counties through which it passed. Three coalfields also lay unworked he argued because of the lack of a rail network to connect them. Baron Ferdinand de Rothschild, then MP for Mid-Bucks, who lived at Waddesdon Manor (also the name of one of the stations) came forward to support Sir Edward, suggesting that the proposed line would bring about a considerable improvement in local communications which had previously been very poor. At Braunston, Colonel Henry Lowndes thought that a direct line would be a useful booster to the local shoe industry and would lead to an increase in the manufacture of other commodities. Nevertheless the population statistics of the previous census in 1881 had shown a drop of almost 100 in the number of villagers living in Braunston at the time.

Public opinion was always a crucial factor in the choice of routes for railways, but in Buckinghamshire the local landlords and prominent figures in Society seem to have been carried away by the idea that the Great Central was intended purely for their use. The traders though welcomed it as a much needed new facility. Two further factors determined the nature and, ultimately, the economic viability of the Great Central. The main occupations of the heart of England were dependent on agriculture and labourers who not only had no reason to travel beyond their village, but even if they had had the inclination, the fares would have been prohibitive.

Active opposition to the new line, largely because of the major engineering works and the disruption involved, came among others, from the artists living in the St. John's Wood colony and from the Marylebone Cricket Club since Lords itself was threatened. A petition was organised at the London end, the protest being directed against the proposed goods depot which was going to engulf some 35 acres of land: The depot was to handle coal, fish, manure and other commodities considered undesirable for such an area. Opposition from the cricketers, however posed the more serious threat and W. G. Grace led the protest in person. Sir Edward reacted without delay because it appeared that his Railway Bill was in grave danger of being rejected by Parliament because of all these objections. He quickly agreed to a compromise. His new line would be hidden by a tunnel and, when the construction work was completed, the pitch at Lords would be restored to its former condition. This was acceptable.

At St. John's Wood however, opposition continued and the case for the Manchester, Sheffield & Lincolnshire Railway was undertaken by Sir Ralph Littler QC. The hearing opened on 17th April 1891 and lasted some 15 days, with Parliament failing to come to any satisfactory decision. Watkin, undaunted by this, pushed to get the Bill read a second time. By the end of March in the following year the Manchester, Sheffield & Lincolnshire Bill was revived and the only objections left now related to the siting of the terminus at Marylebone and the Hotel in Boscabel Gardens. Properties were bought and demolished. The way was now clear.

On 12th April, after a further fortnight's sitting, the committee approved most of the plan for the construction of the Great Central and, after six more days agreed to outstanding matters relating to this. The proposals of the Manchester, Sheffield and Lincolnshire were accepted by the House of Lords. A further hitch occurred when Lord Salisbury dissolved Parliament and it was not until July, when Gladstone took office, that the final consent was given. Royal permission for the Bill was received on 28th March 1893. Meanwhile the Metropolitan had extended its line as far as Aylesbury (1st September 1892). There it connected with the Aylesbury & Buckingham Railway which ended at Verney Junction, the station built for the local landowner and Deputy Chairman of the Aylesbury Railway, Sir Harry Verney.

The Metropolitan intended to extend its line from Quainton Road to Moreton Pinkney, $24\frac{1}{2}$ miles distance, in order to help the MSLR to create a junction with the East and West Junction Railway near the site of the future Woodford Halse station.

The MSLR resolved to build the line the whole way from Beighton in Sheffield to Quainton Road, in Buckinghamshire, where it would meet the Metropolitan at a junction near Upper South Farm, thus abandoning the idea of extending the Metropolitan. Many factors were eventually to frustrate the fulfilment of Watkin's dream of a railway from Manchester to Paris, but for the moment the way was at last clear for Sir Edward to forge ahead with the construction of the London Extension which was an essential part of his grand design.

Like any large-scale civil engineering project, however, the construction of the Great Central Railway involved a number of serious problems. The proposed route about which there was little or no room for manoeuvre, was bound to cause disruption in the already built up towns and cities with the dispossession of some workers from their homes. Moreover it would drive a 'swathe' through the farmlands of Nottinghamshire, Leicestershire, Northamptonshire and Buckinghamshire. Physical problems arising from the nature of the terrain compounded the man-made difficulties. In their turn, canals, rivers, roads and other main lines provided obstacles which had to be crossed or circumvented. Where the route lay through the densely populated cities of Nottingham, Leicester and Rugby, major engineering works in the form of bridges, viaducts and tunnels were necessary to carry the track. But, not the least of many difficulties facing Watkin was the raising of the finance necessary for this ambitious project. The economic situation was indeed very different from that of the earlier years of railway development.

Although British railway companies existed and operated under normal business rules of cost effectiveness and profit, the Great Central company was, from the outset, granted two unusual public privileges, namely corporate status and the ability to acquire property by compulsory purchase. In spite of this however, work on the line was delayed by a number of complicated battles over land purchase. Some of the land in question had a high amenity value as, for example, Lords Cricket ground. Elsewhere when estates were involved, as at Woodford and Catesby there was a conflict of interests since it was felt that the route was likely to disturb the environment and cause annoyance to landowners. Protracted talks ensued and some diversions from the route, or other expedients were necessary.

Eventually, by September 1894, all seven contracts for the line had been placed for a total of £3,132,155, together with a provision of £250,000 for the Channel Tunnel project. Kellett analyses the motives behind railway investment as follows: 'In general the choice of routes, sites and particular operational policies was made privately and according to the ordinary calculations of profitability and investment which prevailed in a laissez-faire economy. The railway companies were business enterprises floated with private capital: in the long run their success and survival depended upon the return which they were able to give their shareholders. The paramount consideration, therefore, in the minds of the projectors and managers of Britain's nineteenth century railway system when making decisions was a relatively simple one – what balance could be expected between the direct private costs and the benefits of the investment? The Victorian entrepreneur was guided by experience and common sense raised to a very high order, not by a systems analysis.'

Judged by such criteria, Watkin's shareholders in the Great Central were involved in a poor investment which paid only 3% on ordinary shares, compared, for example with the high dividend of 60% paid by the Midland Railway Company. Watkin's budget, however, extended in fact to four times his original estimate – a phenomenon of costing far from unknown in modern times – and reached a total of £11,500,000.

In order to bring the Manchester, Sheffield & Lincolnshire to London, authorisation was given for eleven sections of line. The first four covered the route between Annesley and Quainton Road. The fifth was a link line with the London & North Western at Clifton Upon Dunsmore near Rugby. The sixth and seventh contracts were two junctions with the East & West Junction Railway at Woodford Halse. The eighth was for a junction with the Metropolitan at Canfield Place to Marylebone (2 miles 29 chains). The ninth referred to a section where the line was widened between Willesden and Hampstead (1 mile 69 chains). The eleventh and twelfth were for the provision of a coal yard branch of 35 chains in length and a connection to the Inner Circle line west of Baker Street, driving under the site for the Great Central Hotel. Strangely there is no mention of a tenth section, but it is possible that this may have been the line through the Channel Tunnel, or indeed a line across London to join with a route to the south coast, the latter being more likely.

The total cost of the contracts was £3,132,155 divided among seven contractors as detailed opposite.

While researching this book it has been my good fortune to acquire a report or account of the construction of the Great Central which I have reproduced as it has never been published before and actually records the stage by stage construction of the London Extension, just before the line

Contract	Length	Contractor	Cost £
1 Annesley–East Leake	19m 44ch	Logan & Hemmingway, Market Harborough	668,451
2 East Leake–Aylestone	16m 36ch	H. Lovatt, Northampton	548,835
3 Aylestone–Rugby	15m 69ch	Topham, Jones, Railton, Westminster	281,589
4 Rugby–Charwelton	15m 77ch	T. Oliver & Son, Horsham, Sussex	513,308
5 Charwelton–Brackley	12m 32ch	Walter Scott, Newcastle-on-Tyne	
6 Brackley–Quainton Rd.	12m 61ch	Walter Scott, Newcastle-on-Tyne	420,000
7 London–Marylebone (Canfield Place)	1m 71ch	J. T. Firbank, London Bridge	699,972
	94m 70ch		£3,132,155

opened. As well as the Contractors' Journal account, this chapter illustrates some of the fine work by S. W. A. Newton who photographed the line while it was under construction. The latter was born in Leicester at a time when railways were booming but he himself did not become involved with railways until the arrival of the Great Central, which appeared shortly after he had taken on the family photographic business. At this juncture Newton became infatuated with the line as it carved its way through Leicester and after a while he decided to abandon the business in favour of following the Great Central and recording its progress as a travelling photographer.

By the time the Great Central was in the final stages Newton had amassed a substantial number of plates depicting nearly every massive engineering feat as well as recording the social aspects of the navvies who were connected with any task which required brute force. Several shots showed the navvies at leisure, their living quarters, their mission halls and the machinery they worked with. In fact Newton successfully covered as many aspects as possible in the whole six years which must have required considerable effort as there was no railway which linked the places he went to. Undoubtedly without these magnificent pictures which appear in the latter part of this chapter there would have been no record of the construction of England's Last Great Main Line Railway.

It was on the 13th November 1894 that this record actually started with the cutting of the first sod of the London Extension, a ceremony which was performed by the Countess Wharncliffe, wife of the present Chairman of the Railway, in the Alpha Road, St. John's Wood. The following account captures the atmosphere of the scene very well.

'The weather was in striking contrast to that of the previous day, the heavy rain being succeeded by genial sunshine, and the company present which included Sir Edward Watkin, were able to witness the proceedings under the most favourable conditions. As a memento of the occasion the Countess was presented with a large and massive casket of sterling silver weighing 160 ounces. The casket manufactured by Messrs Mappin and Webb was a most elaborate and beautiful work, exhibiting the arms of the Earl and Countess of Wharncliffe, with supporters and motto and the four crests of the family. On the reverse was the ribbon with legends and shields of the five cities and towns with intervening national emblems which form the device of the Great Central Railway Company. The centre circles on the obverse and reverse displayed raised monograms – that of the Earl and Countess on one side and that of the Great Central Railway company on the other'.

For engineering purposes the construction of the line was divided into two sections. The engineers under whose direction the works have been carried out for the northern section of the line were: Mr Edward Parry, M. Inst. C.E. of Nottingham; and for the southern section Sir Douglas Fox V.P. Inst. C.E. and Mr. Francis Fox of Westminster. The northern division extending for some $51\frac{3}{4}$ miles from Annesley to Rugby included Contracts Nos 1, 2 and 3 and comprised the works through the towns of Nottingham, Loughborough and Leicester, while the southern division extending for 42 miles from Rugby to Quainton Road included Contracts Nos 4, 5, 6 and 7 and the two miles from West Hampstead to London (Marylebone).

Contract No. 1 began at Annesley Junction, at a point about a quarter of a mile south of the mouth of the tunnel under the Robin Hood Hills on the Great Central's Annesley line and it extended as far as East Leake. On the immediate right of the point where the London Extension started was a large yard of exchange sidings which had been arranged for the sorting of loaded wagons of coal for the south, and empty wagons from the south to the north so that they might be more conveniently despatched to the collieries for which they were intended. At these exchange or sorting sidings accommodation was originally provided for 1,665 wagons but with the additional land the Company had acquired, it had a plan to develop the site so that a further 2,000 wagons could be stored. In addition to the facilities for dealing with the wagons the sidings that were present before any new construction took place amounted to some 17 miles in length. The yard was fitted out with the most modern equipment consisting of a substantial electric power depot, spacious wagon repairing and engine sheds, turntables, coal stage, balancing table,

Annesley wagon shop.

The engine shed and Annesley exchange sidings.

The coal stage and Annesley exchange sidings.

water tower, numerous water columns, shunters' offices and everything else associated with the development of a first class railway sorting area.

The main job for Annesley was the supply of mineral engines but because of its size and the fact that there was still room for expansion it even took on the role of supplying passenger locomotives when the shed at Arkwright Street closed. High rates were to force the Great Central to cease operations on this site and it is believed that this same factor influenced the company in dropping plans to establish its main locomotive depot at Bulwell which lay close to the site of the carriage sheds at Basford. At one time there was even talk of the carriage sheds being accommodated at Annesley.

To the south of this yard the new railway passed over a private carriage drive to Newstead Abbey, formerly the home of Lord Byron and which lay some two miles distant. Before leaving this part of the line it is necessary to make mention of the fact that very close by lay two active collieries, namely Newstead and Annesley pits which employed between them some two to three thousand men. From this point the railway ran predominantly through an agricultural district, the only features of interest being two important bridges, one over the Great Northern Company's Leen Valley Branch and the other over the Midland Railway to Mansfield. These bridges had to be made of somewhat unusual dimensions in order to provide for possible widenings in case the Channel Tunnel project ever took off. Beyond here the line ran to the west of Hucknall Torkard Church, with which two celebrities, though in different spheres were associated. Lord Byron was buried here, as was the late Ben

Caunt a well known pugilist of his time. Just to the north of Hucknall station there was a huge cutting through magnesium limestone rock, in the removal of which the contractors experienced very considerable difficulty. At the end of this cutting lay Hucknall station and further south still was Hucknall Colliery Company's top pit and the Great Central had constructed a branch connection to give the railway another source of revenue.

Hucknall station with only the goods yard, footbridge and station access canopy to be completed.

The next major place after Hucknall was Bulwell where the railway progressed south over a handsome viaduct of 25 spans and about 390 yards in length, at an average height of 44ft from the ground. In the erection of this structure, which occupied a little over twelve months, six and a half million bricks were used. At Bulwell Station half a mile further along were some important connections with the Great Northern system, those to the north joining the Leen Valley line and those to the south joining the Derbyshire and Staffordshire lines. To the east of Bulwell station lay Bulwell Forest, common land which had recently been purchased by the Nottingham Corporation, and used as a recreation ground by cricketers, footballers and golfers. The golf green was utilised by the Notts Golf Club and the Bulwell Artizans, the latter being one of the few workingmen's golf clubs in England. According to the Contractors' Journal the Notts Golf Club had some redoubtable players and in the 1898 season, a year before the Great Central officially opened, they won all their matches and the closest contest they had fought in all these matches was with their friends and neighbours the Bulwell Artizans when they won by just one hole.

Travelling further south along the railway route, the line had to cross the major obstacle of the valley known as Bagthorpe Hollow, which necessitated the formation of a bank 53ft high and a short cutting 46ft deep. It was from this point that the heavy work of getting through Nottingham commenced. The most northern station in Nottingham was New Basford, where a commodious carriage shed complete with every modern contrivance, and

A very busy looking contractor's yard at Bulwell in Nottinghamshire with a mass of interesting plant visible in the form of mechanical excavators, locomotives, bridge formers, workman's huts and wagons carrying minerals and spoil.

The oil/gas works at New Basford under construction that was to provide power for the carriages which were shedded at the same location.

New Basford station looking towards Sherwood Rise tunnel which is obscured by the steam from a contractor's locomotive in action at the mouth of the tunnel.

an oil gasworks for the lighting of the carriages were erected. Going further down past the island platform station and sidings, the line entered a deep cutting hailing the approach to the northern portal of Sherwood Rise tunnel, which was 662 yards in

Sherwood Rise (Nottingham) tunnel at the northern end, prior to the laying of the main lines. Note the engine on the contractor's track under the northern portal which is built into the sandstone of Sherwood Rise cutting.

length, at the end of which, lay one of the Great Central's few ordinary style stations, namely Carrington. This was designed to be most convenient for businessmen who resided in this popular part of

Carrington station which lay between Sherwood Rise and Mansfield Road tunnels, prior to completion. Carrington was one of the few non island platforms, the main reason being a lack of space.

the town. Carrington in fact was situated between two tunnels, the other being Mansfield Road which started at the southern end of the station platforms and ran for 1,200 yards under Mansfield Road in

Mansfield Road tunnel (north end) Nottingham with Carrington signal cabin to the left of the portal. In this picture the station is far from ready for the surfacing and provision of buildings.

Nottingham. Its construction was unusual in that only the roof was lined with the traditional engineers blue bricks unlike other tunnels. At the southern end the tunnel emerged on the site of Nottingham Victoria Station.

Nottingham Central (later Victoria) station site with a number of the tracks already in place and the formation of the platforms taking shape.

A general view of Nottingham Central station under construction in 1898, a year before the line opened to the public. Central, which was renamed Victoria, was not complete though by March 1899, being first used for passenger trains in 1900.

Victoria was one of the largest stations in the provinces and had a considerable effect on the City during its construction. From end to end it was about 650 yards in length and its average width was about 110 yards. Over 700,000 cubic yards of excavation, principally in the sandstone rock had to be removed to clear out the site before building could begin. Although the spoil was removed in a record time of eight months, the work was rendered unusually difficult by the fact that the excavation had to be commenced from the mouth of the tunnels at both the north and south ends of the station site, and the spoil conveyed through them to the tips on contractors' lines. Near the south end of the site of the Great Central Station the contractors cut through the old city wall, at which point the railway ran through another tunnel which was known as Thurland Street/Victoria Street tunnel. It was only some 400 yards long but the most difficult of the three tunnels to construct due to the fact that it ran in close proximity to the foundations of houses above. Some of these houses in fact had to be carried on timber uprights while the arches were turned underneath.

At the southern end of the tunnel the old Nottingham Guildhall had to be demolished to allow the construction of a long viaduct and a junction starting at Weekday Cross, which in one direction ran to the Grantham line, while in the other it ran on continuous arches to the River Trent. Under the Guildhall where the dungeons were located, the contractors found the bodies of many executed criminals. These were later transported and reburied by the Corporation. Just past this point was Narrow Marsh where a steep decline occurred to the average level of the Trent in this district. From here to the old Recreation Ground, the Nottingham Viaduct extended for three quarters of a mile, crossing a number of streets in the busiest part of the town as well as the Midland Railway, which it bridged by means of a magnificent steel structure which bisected the Midland at right angles.

Leaving the viaduct the railway entered the Nottingham goods yard which was laid out in a manner that could not fail to be appreciated by the Company's customers. An extensive goods shed, worked by hydraulic machinery, roomy goods offices, a complete electric lighting and hydraulic power plant, coal, timber, and stone wharves, cattle

Nottingham goods warehouse was sited north of the River Trent and boasted a fair range of facilities, being one of the most modern freight handling depots of its time.

Nottingham engine shed whose role was later usurped by that at Annesley.

Nottingham carriage shed also fell victim to closure in the early days of the Great Central. Leaving Basford, to the north of Nottingham Victoria, to handle all the coaching stock.

docks, extensive engine and carriage sheds, with all other necessities for equipping a modern depot of a large railway system were provided there. At the south end of the yard the railway crossed the River Trent on an elegant and substantial bridge for four lines of way. It consisted of three spans of 112ft each and was approached on either side by a viaduct. About 2,000 tons of steelwork were used in the construction of the bridge, the work being carried out by Messrs Heenan and Froude of Manchester. From here to the termination of contract No. 1 the railway again ran through purely agricultural terrain, and there was nothing really worthy of note

Trent viaduct (Nottingham).

save for the large cutting at the extreme end, locally known as the big hill at East Leake from which 330,000 cubic yards of spoil were removed in the short period of about nine months. The entire section, extending for nearly 20 miles from Annesley to East Leake involved the excavation and removal of about $3\frac{1}{4}$ million cubic yards of material, the execution of about 200,000 cubic yards of brickwork (in which 70,000,000 bricks were used) and 1,775 lineal yards of viaduct, the construction of 2,253 lineal yards of tunnelling and the laying of 75 miles including sidings, of single line permanent way. There were 75 bridges on this section in which about 10,000 tons of steel and ironwork had been used.

About 2,000 men were employed at one time on the contract, also 20 locomotives, 9 steam navvies, 12 steam cranes, 500 tip wagons, 100 ballast trucks, and 50 horses. The construction of this, the first section of the Great Central Railway Company's extension to London, was carried out by Messrs Logan & Hemmingway, one of the oldest firms of railway contractors in the Kingdom. Two of Mr. Hemmingway's relatives, his grandfather and great uncle, were the contractors for the celebrated tubular bridge over the Menai Straits, which commenced in 1846. The senior partner of the firm Mr John Logan was the MP for the Harborough division of Leicestershire. Mr Charles H. Hemmingway was the resident partner, the manager being Mr Frederick Collins whose name was well known in contracting circles, and whose association with the firm covers a period of nearly 50 years. As well as the works already mentioned Contract No. 1 included two other stations at Nottingham Arkwright Street and Ruddington.

Ruddington station.

Contract No.2 embraced the Loughborough and Leicester sections with a length of $16\frac{1}{2}$ miles and was placed with Mr. Henry Lovatt of Wolverhampton and London. It began in the parish of Normanton-on-Soar with a bank at 19 miles 44 chains, a small but deep cutting of 30,000 cubic yards and a short tunnel of 110 yards in length, commonly known as East Leake or Barnston Tunnel. The tunnel was through rhoetic shale and

Tipping at East Leake, September 1895.

The south end of East Leake tunnel more commonly known as Barnston tunnel in 1897.

required very careful timbering, the excavation being worked out through a heading and the material for a construction lowered down a shaft. There then followed a deep cutting from which

150,000 cubic yards of material were moved. The two cuttings were through similar strata as that of the tunnel, but to the south some exceedingly fine blue lias stone was met with. This was burned into lime and used in the erection of several bridges. Indeed it was of such first class quality that the Barnston Blue Lias Lime Company decided to purchase land and to erect kilns, an enterprise that resulted in good business. There is only one bridge in this cutting and after about three quarters of a mile of bank another cutting was entered which contained about 300,000 cubic yards of red marl. Two bridges spanned this cutting, the one at 24 miles 21 chains, the other being an awkward skew. Near this spot two Roman vases were found along with some copper coins from an earlier settlement.

The next place approached by the contractors was Loughborough Meadows on an embankment 40 to 50ft high which again contained a massive quantity of spoil. In order to expedite the operation a temporary bridge was constructed over the River Soar at roughly 40ft above ground level. At 21 miles 62 chains a viaduct of 11 arches and 160 yards in length had to be built to cross a diverted stream, the foundations, some 25ft deep passing through the gravel into red marl and requiring heavy timbering and powerful pumps to cope with the water encountered. This viaduct as well as the road bridge over the stream and all other bridges and viaducts, were built in brindle facing with red backing in lime mortar. Consequent upon the rapid flow of the water through the openings it was necessary in widening the river to drive 6ft piles at the toe of the slope and to pitch the slopes, and in the neighbourhood of the viaduct to put down a flooring with aprons of cement. The whole of the Loughborough Meadows are, after heavy rain subject to flooding, varying to a depth of from four to six feet and during the construction they more than once sustained considerable damage.

To guard against a repetition it was deemed expedient to drive piling at the toe of the bank slopes and to pitch them with stone. After passing over two or three cattle creeps and occupation bridges (these were bridges built to avoid the need for track level crossings as other railways had), from which a splendid view of Loughborough and the Charnwood Hills could be gained, Nottingham road is reached. This was spanned by a 40ft steel bridge with lattice girders. A few chains further south the Midland Railway main line crossed on a girder bridge of four spans. The next obstacle encountered was a private road bridge and the Loughborough and Leicester Canal which required deep and costly foundations because of the terrain involved. Before reaching the site for Loughborough Central station the Great Central had to accommodate a public road called Borough Street which was carried over the railway on a structure built of steel with brick parapets. Near this bridge lay Loughborough North Cabin strategically placed to give the signalman the best possible vantage point of the station and goods facilities.

Loughborough viaduct under construction.

Aberdeen, ***(Manning Wardle 1014 of 1887, 0–4–0ST) near Loughborough station.***

Turning to the platform this was of the island type and about 500ft long with the usual waiting rooms, booking offices and other passenger requirements. The station itself fell into category 1 which

Right, top: The contractor's yard of Henry Lovatt and Co. at Loughborough.

Right, bottom: **Contractor's locomotive H. Lovatt No. 3** ***(Manning Wardle 1482 of 1900, 0–6–0ST) with three navvies posing for a picture as the engine stands near Loughborough shed.***

H LOVATT No3

Loughborough station yard and platform at a very early stage of construction with most materials on site.

meant that it was an important one and had therefore a more elaborate set up than the simple intermediate stops. A more detailed analysis of this category system will be given in the next chapter explaining which stations are being referred to, under each category. Access to the platform was gained by a large staircase which led down from a new road which was some 60ft in width, which had been constructed on a heavy girder bridge of two spans and on ornamental brick parapets. There were also spacious goods offices, a wharehouse, numerous sidings and approaches together with all the paraphenalia of an important station. Near this station too, at the time of the construction of the new line, a number of manufacturing units and other buildings were being erected in the area.

Proceeding southward a bridge over Beeches Road was the first major structure needed and this consisted of a 40ft girder bridge, while the next was a heavy 70ft structure some 50ft wide to carry the main Leicester Road over the new railway. The last major works before Quorn & Woodhouse entailed the excavation of 150,000 cubic yards of spoil in the form of three cuttings, one of which was so wet that rubble drains with arches had to be installed. At Quorn & Woodhouse which was located in a small cutting, the island platform, like that at Loughborough was again entered from a public road bridge and this station, because it served a small community was only provided with basic facilities. There was a station house and an extensive goods yard for a busy and increasing traffic. The main attraction for gaining the station revenue when it was first planned was that passengers would book

A gang of contractor's navvies trimming the slopes south of Leicester Road.

Woodhouse, Woodhouse Eaves and Quorndon station bridge in 1897. This platform seen in the initial stages of construction was named Quorn & Woodhouse for short at the opening of the London Extension.

there for Charnwood Forest. The next mile ran on an embankment containing about 100,000 cubic yards of earth, relieved by more occupation bridges. Then there was another cutting through 70,000 yards of hard clay and boulders, with a couple of bridges. One of the Leicester Corporation water mains was also carried over it in a steel trough and this was well protected with tan in case of a severe frost.

Progressing further along the route the next feature was the reservoir at Swithland which had to be emptied in order to allow the Great Central to erect two viaducts, one of five spans, the other of

The viaduct over Swithland Water owned by Leicester Corporation Water Works which had to be emptied to allow the construction of the viaducts across the reservoir.

ten to carry the new line across. In fact the journey across the lake boasts some beautiful scenery. On both sides the railway was surrounded by water, well wooded islands and woods beyond, while swans, wild duck and other fowl occupied the water. The shooting and fishing rights were held by Lord Lanesborough who was one of the local landowners. Close to the reservoir on the southern side was Swithland Station yard where sidings were laid principally for the Mountsorrel Granite Company's traffic and for other staple goods. It was the original intention of the company to open a station on this site and provision was made for an island platform but the plan never materialised. Here for the benefit of the Great Central Mr Henry Lovatt constructed a short branch to the Mountsorrel Quarries which was a single line of standard gauge with a 1 in 60 gradient. Half a mile along from Swithland the railway reached its 26th cutting and at the end of this cutting from which some 100,000 yards of sand and marl were taken was the site of Rothley station, being of a similar type to that of other country stations – one island platform approached from an overbridge by a staircase. This was a two arched bridge. There were also ample sidings again with a roadway leading to them.

The next half mile consisted of a heavy embankment containing some 180,000 cubic yards of earthwork, in or about which a stream and occupation road are crossed by a massive brickwork tunnel bridge of 40ft span. Owing to the floods that occur after heavy rainfall it was necessary to invert the stream and build up the side walls, and also to put on the roadway a layer of concrete and piling at the foot of the bank slopes. A subway in the same bank, a bridge over the next cutting with 50,000 cubic yards of excavation, carried to form the embankment, completed the first portion of the contract. The dirt in this cutting was the hardest on the section, consisting of tough boulder clay.

Starting at 29 miles in the parish of Thurcaston the Leicester section (No.2) terminated at 36 miles in Aylestone. More than four miles of this distance were within the borough of Leicester. The works had constantly been exceptionally heavy as will be gathered from the fact that the expenditure on this section alone amounted to £500,000. The first cutting of any importance was the one through Birstall Hill, being nearly a mile and a quarter in length with a maximum depth of 43ft. About 370,000 cubic yards of excavation were next executed, the material being a hard glacial lias boulder clay with some sand and gravel. Birstall Station, situated near the south end of the cutting

Swithland station subway which was bricked up and covered over at the top in anticipation of the construction of another platform, though this never took place due to insufficient demand.

Right, top: The first steam navvy did work in Thurcaston cutting where steep slopes required special drainage runs to be built into them.

Right, bottom: A farmer's accommodation bridge being erected near Cliff Lodge between Rothley and Belgrave & Birstall.

Belgrave & Birstall island platform under construction.

was to be the ticket collecting platform for Leicester. The platform was an island one approached from the main road by a new road and a bridge over the cutting in true Great Central style. We next come to a bank 46ft deep and containing 160,000 yards of earth. At its deepest part the Thurcaston Road is crossed by a heavy arched bridge 80ft 7in long and with a height of $18\frac{1}{2}$ft. The skew span is 48ft 10in, that on the square being 40ft. The course of the face of the arch were laid at right angles to the heading spirals and in the centre parallel to the abutments with intermediate curved portions.

The completed Thurcaston to Rothley and Mountsorrel Road bridge in 1897 which formed the station access to Rothley platform.

There were altogether nine arches on the Leicester section constructed on this principle, this being the largest. At 32 miles the coal yard at Abbey Lane was designed for the convenience of large suburban traffic which was rapidly growing on this side of the town. Half a mile further on Leicester itself was reached, and at Harrison Street a viaduct began which ran for more than a mile over some of the principal rights of way. In the centre of this viaduct Leicester Central station was built. The rails were at an average height of 20ft above the level of the streets which were crossed by girder spans, the headroom being insufficient for the employment of arches. For the intermediate portions arches were adopted everywhere, the spans being approximately 30ft and the rise invariably 10ft. Abbey Gate bridge had a span of 65ft 9in, the weight of the steelwork being 130 tons, while that over the River Soar at 32 miles 58 chains had two spans of 80ft 1½in and 81ft 0½in, the steelwork weighing 300tons. The pier in the centre of the river which is shown in one of the pictures, was constructed within a cofferdam of single sheet piling.

The bridge over the Leicester Canal at 32 miles 72 chains has a span of 69ft 10¾in, with a weight of 129tons and for this structure the fish bellied type of girder was adopted in order to obtain the necessary

Leicester North Goods Cabin.

Leicester Station retaining walls near Northgate Lane in the process of erection in 1897.

headroom for the road it passed over. Northgate Street bridge, which had a span of 107ft and 257tons of steelwork, is fan shaped in plan as it was here that the lines began to open out for Leicester Central station. The bridge at All Saints' Road, the heaviest of the three that had to be constructed beneath the station had a span of 43ft, the length of its abutments being 194ft and its weight being a massive 254 tons. At this point between Friar's Causeway and Jewry Lane was a fine example of a Roman Pavement in a splendid state of preservation. It had always been jealously guarded by the municipal authorities so that in bridging over the site and making the approaches thereto, the Great Central had had to incur a considerable expenditure. At 33

Leicester station platforms and glass light covering for the Roman Pavement which was unearthed during the excavations for the platforms.

Erection of the Massive viaduct over St Augustines Street.

The Navvy Mission Room built into the fifth arch of the splendid brick viaduct at Duns Lane which had its own chaplain and missionary.

The interior of the Mission Room at Duns Lane.

miles 32 chains the River Soar was again crossed, this time with a single span bridge, while other magnificent engineering feats lay at Braunstone gate and over the old course of the River Soar.

The island platform of Leicester Central station was 1,300ft long by 80ft wide and at each end in between the main platforms two bays were constructed for local services. A large portion of the roof was covered with a substantial canopy supported on steel columns. Three blocks of buildings were erected, containing waiting rooms, usual offices, dining and refreshment rooms in the centre of operations. Beneath these were arranged kitchens, cellars, stores and offices. The station was entered from street level by means of subways from both sides. These were lined with buff glazed bricks, and after meeting at the centre, the rise to the platform was made on York stone steps. A lift, worked by hydraulic power, was provided for luggage. The booking and parcels offices, cab stand etc were situated at street level on the eastern side of the station. The buildings were faced with pressed red Heather bricks and with buff terra cotta from Hathern. The style may be described as a classical railway renaissance. No expense was spared to secure a handsome and bright appearance, the materials and workmanship being the very best. Along each side of the island platform there were four roads, whilst there was a carriage dock, fish and milk stage, engine turntable, coal stage and standing sidings which were located at the north east end.

Returning to the main line after leaving the viaduct the new goods yard was traversed on an embankment, four roads being laid in order to keep the goods and passenger traffic apart. Also at this point a river diversion was needed and three massive bridges were erected to carry the new roads to different parts of the goods yard. The viaduct, which was a conspicuous object in the middle distance, carried a new 50ft street over the end of the goods yard, while beneath its central openings lay the locomotive shed, wagon repairing shops etc. In securing a site of over 40 acres in the heart of the town the Great Central was extremely fortunate. The land had previously served as a recreation ground, and in its new role had been laid out and equipped in the most modern and up to date manner. Coal sidings timber and stone yards and cattle pens occupied the east side. The central portion was devoted chiefly to the goods warehouse with its approaches and sidings. In fact there was even provision made for a possible extension if ever the traffic flow warranted it.

The portion of the yard to the west of the main line was occupied by a long carriage shed, which had four roads and these were provided with a succession of troughs to supply hot and cold water for washing purposes. Other large buildings in the goods yard were a fine block of offices in red brick and Derbyshire stone, a hydraulic engine shed with five large steam boilers and five dynamos with high speed engines, and a wagon repair shop powered by electricity. Extra facilities such as a boiler house, carriage repair shop, two weighbridges, offices for yardman, a goliath crane and a five ton crane were all provided. Adjoining the goods yard was a locomotive depot for which some 30 acres of land were acquired, some of which it was hoped would be needed for extensions at a later date. The locomotive shed was over 300ft long and 97ft wide. It had four roads with pits and was built to stable engines on express runs between Marylebone and Leicester and also to provide relief engines for those trains being worked on to Manchester. Erections alongside the shed comprised of an engine room, machine shop, smiths' shops, stores, mess room and offices. Also provided in the locomotive yard were a coal stage, shear legs, a balancing table and turntable.

Leaving the yard the main line proceeded through a 20ft cutting for $\frac{1}{4}$ mile until it reached Aylestone Viaduct – a four arched brick structure

An accommodation bridge was even provided for the footpaths along the London Extension and this one, which bridges the Aylestone to Narborough footpath, stands in isolation waiting for the embankments to be provided on either side!

built on a skew pattern. Next, the line crossed the Soar Valley on a bank $\frac{3}{4}$ mile long, the only feature of interest here being a girder bridge where the River Soar incommoded the railway for the sixth time on its course through Leicester. Half a mile beyond this bridge was the end of the second contract which was completed in a record two and a half years, during which time 1,000 to 1,500 men were regularly employed. Upwards of 6,000 tons of steelwork, 75 million bricks and 120,000 cubic yards of Ashlar were used in the building of this section. The plant included 16 locomotives, 6 steam navvies, 15 cranes, 25 portables and boilers, 14 mortar mills and 12 steam pumps along with 40 miles of rail, 500 tip wagons, 100 ballast wagons and 50 horses.

Contract No.3 included that portion of the railway between Aylestone and Rugby, a distance of $15\frac{1}{3}$ miles. The work generally was of a heavy nature, and on leaving Aylestone the cuttings for a mile or so cut through red marl. Near Whetstone the railway crossed the Union Canal by a girder bridge and the River Sence by a viaduct of 13 spans faced with blue brick. Soon after leaving the viaduct the line crossed the branch railway of the London & North Western, from Nuneaton to Leicester, by a girder bridge of 86ft span. At Whetstone some four miles from the centre of Leicester, the Great Central provided a commodious station together with the usual buildings, as well as a respectable goods yard to woo local traffic. At Whetstone too there commenced an embankment three miles in length, the gradient from here ascending 210ft in 7 miles. Six miles from Leicester the line crossed over the Midland branch line from Leicester to Rugby, and then it entered a cutting over which the Leicester and Rugby main road was carried on a three arch skew bridge. From this point to the end of the contract the cuttings are in lias formation and drift. At the afore mentioned spot the contractors established their yard and erected a locomotive shed, offices, repair shops and workmen's dwellings. The second station on this portion of the line was at Ashby Magna where the necessary sidings cattle docks, platforms and the like were provided. This station was at first designed to serve Dunton Bassett, Cosby, Broughton Astley and Gilmorton.

A little to the south of Ashby Magna was a cutting and a tunnel under a public road. From

The join of Contracts Nos 2 and 3 near Aylestone.

Dunton Bassett tunnel which lay just to the south of Ashby Magna island platform and goods yard.

Ashby Magna to Lutterworth the country is of an undulating nature, and at Gilmorton 12 miles from Leicester, the highest point on this division of the line was reached, the surface level being 455ft above sea level. From Gilmorton to Lutterworth the line descends by a gradient of 1 in 176. At Lutterworth there was a convenient little station complete with all the facilities as well as having a station master's house. From here to the end of the contract the line crossed no major obstacles and in fact this contract, performed by Messrs Topham Jones and Railton was probably the lightest of the seven.

Contract No.4 which was entrusted to Thomas Oliver and comprised a length of 15 miles 60 chains. It commenced at the point where the northern section ended and terminated at 67 miles 20 chains on the main line. There was also a short branch railway of 35 chains connecting the main line at Woodford station with the East & West Junction Railway from Towcester to Stratford upon Avon. The contract had involved some heavy work, including the construction of 5 viaducts, 46 bridges, 32 culverts, 12 cuttings, 13 embankments, 6 sets of sidings, a junction and a tunnel nearly $1\frac{3}{4}$ miles long. The whole of the brickwork in the bridges, built on concrete foundations was composed of local bricks with Staffordshire brindle brick facings. All the girders were of steel. The bottom ballast was mostly hand laid granite, and the top ballast was broken slag or gravel.

In commencing their work the contractors had an extremely difficult task to do. Over the Oxford Canal, to whence the railway was carried on a 30ft embankment, a massive viaduct had to be constructed without interference to the traffic, the company being under an agreement stipulating a fine if barges were held up. To guard against an escape of water from the canal in the event of any settlement which might have occurred during the sinking of the cylinders through the embankment special precautions had to be adopted. The difficulty was overcome by impounding the water of the canal for some distance in a steel trough, sufficiently large enough for boats to pass through. The canal was crossed by a steel girder span of 110ft resting on steel cylinders sunk through the canal bank into the old surface, and three spans of 91ft resting on brick piers and abutments. At the end of the viaduct was an embankment about a quarter of a mile long and this bank joined the viaduct over the London & North Western Railway near Rugby, and contained 300,000 cubic yards of material. The spans of this

Lutterworth station.

The 'Birdcage Bridge' which spanned the London & North Western's West Coast Main Line, just to the north of Rugby Central station.

structure had to be erected on a staging and lowered into place above the sidings and main lines of the LNWR. Remarkably this was achieved without any disruption to traffic.

After leaving the viaduct was the Rugby cutting, just under two miles in length. It had involved the excavation and removal of nearly 1,500,000 yards of clay, silt and dirty gravel. A large volume of water had to be tapped and a deep heading drain had to be made on both sides of the line for a long distance before the excavation could be proceeded with, the water being pumped out at all the low points in this drain by steam pumps working day and night.

Six bridges crossed over this cutting, the principal one being at Hillmorton Road forming the approach to the station and another composed of three arches which carried the Dunchurch Road. From the end of the Rugby cutting the railway for the next two miles was on an embankment containing 550,000 cubic yards of earthwork. Then for a further two miles the railway ran along ordinary cuttings and embankments. At Braunston the River Leam was crossed by a twelve arch viaduct, the embankments at the end of which were tolerably heavy and contained some 350,000 yards of spoil. Within the same mile the railway again passed over the Oxford Canal and over the Daventry and Leamington branch of the North Western by ordinary girder bridges.

Heavy cuttings and embankments with the usual under and over bridges, carried the lie as far as Staverton Viaduct, which was built with nine arches. The next viaduct approached was that at Catesby after a short cutting. Half a mile further on Catesby Tunnel was approached and it commenced

Staverton viaduct.

Catesby viaduct.

near the house of that name, (where the Gunpowder Plot was supposed to have been hatched) and it was roughly 3,000 yards long. It was lined from end to end with brickwork and had inverts, and was faced with Staffordshire blue bricks. Some 30 million bricks were used in its construction. The tunnel was worked by driving a bottom heading for about 440 yards from the north face and putting up five 'breakups' from the bottom heading. The ground at this end was very heavy, and owing to an agreement with the landowner no shafts could be sunk or the surface be broken for a distance of 500 yards from the north face. The tunnel therefore had to be driven the full size until there was only about 4ft from the top of the brickwork to the surface of the ground. In this operation great care was taken to ensure that the tunnel was properly supported during digging.

At the end of the bottom heading a shaft was sunk, and from this shaft to the south face an overland route was laid. Nine shafts were sunk and the

Catesby tunnel.

tunnel driven full size from both ends at each shaft until a junction had been made throughout. Water was tapped at all shafts, and as the gradient was falling at 1 in 175 from south to north, special means had to be used to pump the water from the shafts. Very rapid progress was made in the cutting of the tunnel and it was completed in 25 months. Bearing in mind the length, it was probably the fastest time for completion of such a feat. Emerging from the tunnel the next major area was the complex of Charwelton station which was provided with the typical country station buildings, but because it was an important village for ironstone mining the Great Central provided extra sidings.

Of all the features in the section, that at Woodford was probably the most demanding as here vast amounts of spoil had to be heaped up to provide accommodation for the locomotive depot, station, goods yards and all the other facilities associated with a major depot. The locomotive shed was built to house 30 engines with provision for 30 more, while a vast complex of wagon and carriage shops, turntable, sidings, coal stage and water towers was provided. In between here and the station lay the massive embankment which contained some 430,000 yards of spoil. Upon leaving Woodford station, which was provided with its own goods sidings and cattle landings, the point of the junction with the East & West Junction Railway was reached. Three other stations on this section were Rugby, Willoughby and Charwelton which has already been described.

Contracts Nos 5 and 6 covered $24\frac{1}{2}$ miles of main line, a loop of three quarters of a mile long at Woodford forming the south junction with the East & West Junction Railway, and a branch of $8\frac{1}{4}$ miles from near Eydon to the Great Western main line to the north of Banbury station. In all $33\frac{1}{2}$ miles of works existed, which were placed with Messrs Walter Scott and Co. of Newcastle and Westminster. In connection with the construction of the Banbury branch, which passed through the villages of Culworth, Thorpe Mandeville and Chacombe, there were no special engineering features other than a large cutting, containing 900,000 cubic yards of excavation at the second mentioned place, and an embankment $2\frac{3}{4}$ miles long which ran through Chacombe Meadows to the junction with the Great Western Railway. From Woodford (contract No.5) the Great Central Railway, after passing under the EWJR proceeded to Helmdon where it crossed the Northampton & Banbury Railway on a nine arched viaduct, which passed close by the village of Sulgrave, the ancestral home of George Washington.

At Brackley, where the barons are alleged to

Brackley viaduct.

have rested whilst they negotiated with King John at Oxford at the beginning of the Magna Carta struggle, there was a viaduct consisting of 21 arches and 2 girder spans. The foundations of this edifice gave considerable trouble to the contractors because of the nature of the ground. Sliding beds of clay made the removal of the three southern arches necessary, one being rebuilt as an abutment while the other two were reinforcing girders. A heavy concrete retaining wall had also to be built to fortify the embankment. The total quantity of earthworks in the embankments of this section amounted to upwards of three million cubic yards, those at Eydon, Sulgrave and Brackley being each responsible for the removal of 500,000 yards, while cuttings at Barrow Hill, Radstone, and Brackley involved the extraction of similar quantities.

After leaving Brackley, where a large station was provided the line crossed over the Ouse Valley on another large embankment, the river itself being carried through it in a 20ft culvert. The section finished in the Parish of Mixbury, Oxon $1\frac{1}{2}$ miles to the south of Brackley where the line crossed over the Banbury branch of the London & North Western, which was effected by means of a large girder bridge. Contract No.6 which started here was of a much lighter nature, the only heavy work being the excavation of cuttings at Mixbury, Finmere, Chetwode and Steeple Claydon, amounting to some 900,000 yards of spoil.

Near Steeple Claydon the Great Central was carried over the Oxford branch of the North Western by a plate girder bridge similar to that used in bridging the Banbury branch. The total quantity of earthwork on these contracts exceeded 5,000,000 cubic yards, the principal part of which was in blue lias clay, although limestone rock in quantities was met with at Helmdon, Brackley, Mixbury and Thorpe Mandeville. The maximum number of men engaged at the busiest time was 2,300 in addition to which the contractors employed a considerable amount of plant. Many of the materials were brought in from other parts of the country but the red bricks which were used in viaduct and bridge construction, were locally obtained or in fact made in temporary brick yards established by the contractors from where they were transported to site by contractors' railway lines.

One of the problems encountered by the contractors was the distinct lack of housing for the workmen, so several colonies of temporary huts were erected along the line, at four of which navvy mission rooms were also provided. As these latter were used as reading and recreation rooms, night schools etc, as well as for religious services they were very much appreciated by the men and their families.

Calvert station.

Returning to the stations on this section from Woodford, these were Culworth, Helmdon, Brackley, Finmere and Calvert which was the last on the London Extension before the junction with the Metropolitan at Quainton Road. The title Calvert was borne by none of the local villages but it was the Verney family name and the late Sir Harry Verney, a great railway figure, made it a condition that on agreeing to the passage of the railway through his estate that the station should be so named. The booking offices, waiting rooms, and the stationmasters' houses etc were erected by a firm of Rugby engineers.

Despite the strenuous opposition of the Metropolitan, the Great Central company obtained an act in 1898 to build a short line to link the GC main line with that of the Great Western, from Grendon Underwood to Ashendon and then, at the London end to build a line from the Metropolitan at Neasden to High Wycombe. This enabled the Great Central to avoid the need to run over the already congested Metropolitan and also meant that the possibility of running into Paddington as well as Marylebone was there.

Marylebone terminus and the two miles of track from Finchley Road junction formed the seventh contract. The first part consisted of nearly $1\frac{1}{2}$ miles of tunnel or covered way. The track comprised of a double set of metals but with an augmented traffic in mind the Great Central provided and prepared a second tunnel for an extra pair of rails. The three tunnels under Lords Cricket Ground at the time of construction had no less than seven pairs of rails under them but only three of these were used for through traffic. On the bridge over the Regent's Canal close to St. John's Wood station there were 11 distinct lines. One of them on the extreme right, curving off along the bank of the canal to a wharf which was fully equipped with

The last bridge on the London Extension marking the boundary between the Great Central and the Metropolitan Railway. The latter branched off from Verney Junction, north of Quainton Road where the London Extension metals started.

cranes and other appliances for the rapid interchange of traffic with barges. Another pair of rails ran down a decline so as to pass under Grove Road to the coal depot which was entered from Carlisle Street in convenient proximity to the Edgware Road. Although fairly extensive the depot soon proved to be too small once the line had opened, indeed it did receive a considerable extension because of this.

Still further round and running close to the passenger lines was the goods depot which was provided with an immense warehouse in which were the latest hydraulic and electrical machines, to aid the transporting of merchandise. Near the warehouse the Great Central created a new thoroughfare called Rossmore Road. Moving south in between this new road and the Marylebone Road lay the terminus itself which lay at street level, occupying the former Boscabel Gardens. Access was made by a new road, Great Central Street on the eastern side of which was an independent affair, the Great Central Hotel.

The terminus had three clerestory roofs and covered two island platforms and five sets of rails and was built in a plain style. The platforms were 1,000ft long and the whole station, including the goods yard took up some 70 acres. Returning to the operations area, opposite the goods yard lay the carriage sidings and sheds along with a horse dock, milk and fish platforms though there was no locomotive shed as this was provided at Neasden. With the completion of Marylebone and indeed the end of some five years work, the Railway was ready to open to passengers in 1899. The line had in fact been in use the previous year for running coal trains but the eager passenger had to wait until March 1899 before they could experience the fruits of the navvies' labours. By the time the railway opened the only structure unfinished was that at Nottingham Victoria or Central as it then was and this station was commissioned on 24th May 1900.

Shortly after opening, a couple of passengers admire one of the ornamental gardens inside Marylebone station, London. Marylebone was always noted for its pleasant light and airy atmosphere similar to that of a gentleman's club. Note the 'finger' sign to the right of the plant pot under the refreshment room sign, advertising the range of destinations available from Marylebone's platforms.

Chapter 2

All Stations to Nottingham

By 1898, following almost ten years of preparation, the most exciting stage for the instigators of the Great Central London Extension project was about to take place – the running of the very first trains between Annesley and Marylebone. These consisted of coal trains which began running in July 1898 and were used until it was thought that the earthworks had settled sufficiently to bear the increased weights that passenger workings would bring about.

November of that same year saw the first train of passenger stock reach Marylebone, which was brought from Manchester in the charge of Pollitt Class 11a No.268. This train did not carry passengers and was used for evaluating the route and examining the ride quality. It was to be another four months before the line actually opened to passenger traffic and on this splendid occasion the whole of the new station concourse was laid out for a sumptuous banquet and decked out with flags and bunting to herald the opening of the last main line to be built in England.

Contemporary photographs, some of which are included in the following pages, show all the guests wearing their top hats actually at the dinner table! The epicurean meal was more than worthy of the occasion and once completed the only function left to perform was the setting off of the first train, which was carried out by the Chairman of the Great Central.

The name Great Central was bestowed on the railway as it was thought that the Manchester, Sheffield, Lincoln and London Railway was too long winded, and as the route of the line ran down the centre or spine of England, so the name 'Great Central' came into being. The Chairman of the GCR, the Right Honourable C. T. Ritchie, who was president of the Board of Trade was handed a large silver casket at the ceremony which contained a vellum scroll depicting the course of the route. In his speech Ritchie observed quite rightly that there was little probability that any further great trunk railway route would ever reach London again. Though the dream of Sir Edward Watkin had been

partially achieved he was in ill health by the time of the opening and attended the ceremony in a bath chair. Sir Edward and all the other guests were arranged on platform 2 in a group, while Ritchie and his wife, the directors and heads of departments were assembled on a dais. Ritchie pulled a silver lever to admit steam into the cylinders of locomotive No.861 and the train moved slowly out of Marylebone to the cheers of the assembled guests and the general public lining the route beyond.

Services for the public were instituted on the 15th March 1899 and the first three trains carried 4, 14 and 34 passengers respectively. These initial figures were unfortunately all too significant, since the low density of traffic underlined the existing competition that the line was to face from the other major established companies with routes into London. After the first trains the service then consisted of five trains daily at 5.15am, 9.15am, 1.15pm, 5.15pm, and 10.15pm. The scheduled speeds were low at first as the earthworks of the line were still in the process of consolidation and service slacks were numerous. The train service was subjected to

Right away – a top hatted crowd has been assembled at Marylebone to witness the send-off of the first train on 9th March 1899, hauled by locomotive No.861. Note the period plate camera erected on a tall tripod by the tender, and the silver lever on the table by the cab. This was ceremonially pulled by Colonel Ritchie to operate the locomotive regulator and thus start this historic train on its journey. ***(Leicester Museums)***

In Great Central days a group of retired railway employees pose for the photographer by the engine on a Marylebone-Manchester train standing in platform 4. From left to right the men are Thomas Wellan, William Kaye, William Farrand, George Stocks, William Crossley, John Berkinshaw, William Mawson and wife, John Preedy, John Gott and Joshua Glover.
(G.C. Museum, Loughborough)

a great deal of revision during the following years but the timetable shown (the earliest one which exists) shows what was provided once the railway had finally settled in earnest to work. At this time Nottingham Victoria was incomplete and trains called at Arkwright Street at the south end and

Timetable of G.C. Expresses 1899

	News Paper	B.C.	R.C.	B.C.	B.C.	B.C.	R.C.	SUNDAY		B.C	R.C.
Marylebone	5.15	8.00	10.00	12.00	1.30	4.00	6.15		Dep	11.00	5.00
Leicester (Central)	7.15NS	10.24	12.10NS	2.17	3.39NS	6.24	8.25NS		Arr	1.10NS	7.22
Nottingham (Arkwright St)	7.52	11.00	12.45	2.53	4.12	7.00	9.00		Arr	1.47	7.56
Sheffield	8.47	11.54	1.39	3.48	5.06	7.54	9.56		Arr	2.43	8.50
Manchester (London Rd)	10.00	1.05	2.50	5.00	6.15	9.05	11.05		Arr	3.55	10.00
	B.C.	R.C.	B.C.		B.C.	R.C.	B.C.			B.C.	R.C.
Manchester (London Rd)	8.10	10.00	12.00	2.00	3.30	5.00	7.30		Dep	12.30	5.30
Sheffield	9.18	11.07	1.00	3.07	4.37	6.07	8.40		Dep	1.40	6.42
Nottingham	10.13	12.01	2.06	4.01	5.31	7.01	9.36		Dep	2.36	7.40
Leicester	10.50	12.50	2.55	4.35	6.07	7.35	10.12		Dep	3.12	8.18
Marylebone	1.25	3.10NS	5.30	6.45NS	8.40	9.45	12.45		Arr	5.25	10.35

B.C. Buffet Car R.C. Restuarant Car N.S. Non Stop

Carrington at the north end though it was only possible for trains to stop at these places for a very short time due to the fact that there were no passing places available. Most of the buffet car trains also called at Aylesbury and Harrow on the Hill. All of these expresses were vestibuled throughout.

In 1900 the opening of two more parts of the London Extension occurred, namely the Woodford to Banbury link which meant that traffic for the Great Western could now be exchanged at Banbury instead of at Aylesbury. Three shuttle trains were provided between the two centres and there was also a new service introduced between Leicester and Oxford. In fact this little section of line was to be responsible for a massive increase in freight traffic on the Great Central.

The other part of the Great Central to open was Nottingham Victoria station which had not been complete when passenger trains first started running. The opening of Victoria took place on 24th May 1900 without any formal ceremony save for the striking of some medallions by a local firm which depicted the station. Until this time the station had been called 'Joint' because it served both the Great Northern and Great Central companies. The Great Northern gained access via a line which met the London Extension near Arkwright Street at a junction called Weekday Cross. There was also a link to the north of Basford at Bagthorpe. The Great Northern referred to the station as 'Joint' while the Great Central wanted to call it 'Central', but it was to be the date of opening and the Mayor of Nottingham who was to settle the dispute over the name. The 24th May was Queen Victoria's birthday and a suggestion was put forward, that was duly adopted that the station should be called Victoria. The station sat in a large cutting dug out of the Nottingham rock and was largely out of sight. The main buildings stood on Mansfield Road and there was a hotel provided for weary travellers which still survives today. The ticket office and general offices were marked out by a striking clock tower while the building itself was in red brick and styled with 'Dutch type' gables making it a very impressive structure. An awning covered the forecourt and cab pick up point and from this forecourt there was a roadway which ran down into the main station area.

Nottingham Victoria station with a Great Northern Railway 2-4-0 in platform 7. The station was initially referred to as Nottingham Joint Station. *(G. C. Museum, Loughborough)*

A later, exterior view of Nottingham Victoria station.
(G. C. Museum, Loughborough)

The buildings on the platforms were built from local Darley Dale stone and Nottingham bricks. The large blocks of buildings on the platform were elegant in design and beautifully faced with light yellow glazed bricks and the whole roof was covered with an immense mass of ironwork which supported the large glass canopy. Although the station was below the ground level the canopy with its large surface area of glass helped to give the concourse a light airy effect similar to that at Marylebone and some even said that it had the ambiance of a gentletlemans' club. The platform was in the usual island formation though a break with tradition had been incorporated in the design as there were three roads running through the centre of the formation.

Victoria, because of its location, was an immense station which could serve a number of destinations from its bay platforms as well as the through services. Each platform contained lavish facilities such as refreshment rooms, waiting rooms and stores etc. One notable feature was the generous provision of cloakrooms due to the fact that not all trains had toilet facilities or corridors at the time when the station was constructed. The station also possessed hydraulic lifts for each platform and there was a massive clock mounted on the centre of the footbridge over the main tracks. Altogether there were twelve platforms on the two islands: Nos 1 and 4 served northbound trains while Nos 7 and 10 served southbound workings. Either side of platform 1 and 10 there were two passing loops where goods trains could be halted whilst fast passenger or freight workings made their dash to the north through Mansfield Road tunnel. Storage for carriages used on local trains could be stabled on a long central siding which ran between platforms 4 and 7.

For local trains there were eight bay platforms, four at each end and invariably three of these were used for the storage of parcels vans. The other five were used for trains to Grantham, Derby, Sheffield, Chesterfield, Newark, Mansfield, Pinxton, Rugby, Basford, Stafford, Leicester and Burton upon Trent. To control all these different workings some four signalboxes were provided, one each at the north and south end of the platform with one in the centre of each island platform. At either end also the station was provided with a few sidings and a turntable, although it had the unfortunate distinction of being too short for some of the larger engines, and it was not uncommon to see a tender hanging off the back of the railbed.

Turning to the long distance trains to which

the station played host, these were the Manchester to Marylebone workings and the rather less well patronised Bradford, Huddersfield and Halifax train, and numerous cross country workings such as the York to Bournemouth and Newcastle-Bournemouth trains. The Newcastle to Swansea train which called here was unofficially dubbed the 'Port to Ports'. The train from Newcastle passed through Nottingham at 1.55pm while the one in the opposite direction passed through at 2.12pm. The cross country workings were always a source of great variety as far as rolling stock was concerned as it was a regular thing for the chocolate and cream of the Great Western and the green of the Southern to be seen on Great Central metals. One of the most fascinating workings which ran over the London Extension was that from Aberdeen which was done by means of a through coach which covered the distance of 792 miles. The service was inaugurated on 3rd October 1921 and the table shows the press runs before the service was inaugurated. The coach was attached to a rake of passenger and parcels vehicles and as one can see, called at Victoria 8.38pm going south and 9.15pm going north.

Nottingham Victoria during summer Saturdays catered for holiday trains which ran to Mablethorpe, Skegness, Cleethorpes, Great Yarmouth and Scarborough and other resorts. These services were always very popular, mainly on account of the cheapness of fares. Other trains which served Victoria included workers' trains which only ran short distances, such as the one from New Basford to Arkwright Street which ran during weekday lunch hours. There were also steam railcar services to Radcliffe-on-Trent and Ilkeston for the same reason. In all, Nottingham Victoria was a busy

ABERDEEN – PENZANCE THROUGH SERVICE

Inaugurated 3rd October 1921. Press Journeys. Southward 29th September 1921.
Northward 30th September 1921.

Aberdeen	Dep	9.45 am	Penzance	Dep	11.00 am
Edinburgh		1.30 pm	St. Ives		10.25
Newcastle		4.21	Falmouth		11.20
York		6.25	Newquay		11.15
Sheffield		7.36	Plymouth		2.00 pm
Leicester		9.04	Torquay		2.23
Banbury	Arr	10.00	Exeter		3.27
Oxford		10.33	Swindon		6.15
Swindon		11.30	Oxford		7.07
Exeter		2.46 am	Banbury		7.42
Torquay		3.50	Leicester	Arr	8.38
Plymouth		4.25	Sheffield		10.04
Newquay		7.25	York		11.15
Falmouth		7.23	Newcastle		1.20 am
St. Ives		8.36	Edinburgh		4.05
Penzance		7.40	Aberdeen		7.40

Aberdeen – Penzance – Brake Composite Coach – NBR and GWR alternately
York–Penzance – Brake Third (GCR)
York–Swindon – Brake First, Composite Dining. Open Third – NER and GWR alternately

Southbound	Through coach attached to Aberdeen–King's Cross express and detached at York Through coaches attached to 10.00 pm sleeping car express Paddington to Penzance (via Bristol)
Northbound	Through coaches attached to 11.00 am express Penzance to Paddington – detached at Westbury Through coaches attached to 7.00 pm ex King's Cross sleeping car express at York

New mileage	–	Southbound. York to Swindon 204 miles Northbound. Westbury to York 220 miles
Total mileage	–	Aberdeen–Penzance 785 miles
Nameboard	–	"Aberdeen and Penzance. Via Edinburgh, York, Sheffield, Leicester, Swindon and Plymouth."

station as far as passenger trains were concerned and when one included the numerous freight workings there seemed to be a continuous bustle of traffic in the form of coal trains from the Notts and Derby coalfields, iron for Stanton and Staveley and steel products for the Sheffield area. On top of these there were also the fish trains from Hull and Grimsby.

Leaving Nottingham Victoria there were four stations before the next large centre was reached. The first was at Arkwright Street situated at the south end of the City in a residential and industrial area while the second and third, Ruddington and East Leake were out in the countryside and served small villages. The fourth which was a platform called Rushcliffe Halt, was not brought into use until 1911. The design was similar to that of Arkwright Street and Carrington which were also built on the conventional basis due to lack of available space.

Stations fell into four main categories as far as design was concerned and varied according to the size of the community they served. The first category were those built on the usual method ie Carrington, Arkwright Street and Rushcliffe Halt. The second type referred to the larger stations which had large island platforms with more buildings than usual, such as at Nottingham, Loughborough, Leicester, Rugby, Brackley and Woodford Halse. The third category refers to those country stations which were built on the island principle and had access by means of a staircase up to the platform as at New Basford, East Leake, Whetstone, Lutterworth, Braunston and Willoughby and Finmere while the fourth and final category encompassed those island platforms where access was made by a staircase down to the platform as at Hucknall, Ruddington, Quorn & Woodhouse, Rothley, Belgrave & Birstall, Ashby Magna, Culworth, Helmdon and Calvert. Although there were four specific designs for the platforms no one station was like the next as each had its own peculiarity, such as there was an extra building at Birstall, Woodford Halse had a footbridge, Brackley and Rugby had their ticket offices as road level instead of on the platform. These small stations were very spartan and functional containing only the minimum number of buildings unlike their larger counterparts of category two, which were provided with lavish facilities.

Loughborough Central was the next port of

No.44848, Class 5 4–6–0 on Friday 10th July 1964 leaving the typical Great Central island platform of East Leake with the 5.15pm ex Nottingham Victoria to Marylebone semi-fast service. Note the single lens aspect colour light signal by the 'down' line. *(T. Boustead)*

call of any significance though the town was already served by the Midland on the St. Pancras route and Derby Road which was at the end of a very sinuous line from Derby. Because of Loughborough's vast size Central was provided with extensive facilities as far as the station was concerned. There was a large building at road level which contained the booking office, parcels office and luggage room, while down on the platform there was a refreshment room, toilets and ladies' and gentlemen's waiting rooms as well as a general waiting room. All these buildings were covered by a continuous canopy which ran the length of both platforms. For freight Loughborough boasted more than adequate accommodation and this was situated on the 'down' side of the station. One long siding however did exist on the 'up' side. There was a siding which served a building yard off Queen's Road which was linked to the main goods yard which was sited closer to the station itself. Lying off Warren Street was a wharf and cattle pens which were served by two sidings while two roads ran into a goods shed in front of which lay two short

storage sidings. Goods traffic consisted mainly of timber, coal and materials for local industry such as hosiery, brickmaking and engineering.

As well as the freight side, Loughborough was served by an adequate number of passenger services right from the time when the line opened and in the summer of 1903 the Great Central provided five through trains to Marylebone, while it was also possible to change at Leicester and reach London by another five trains. From Loughborough to Manchester there was a service of five trains to Manchester which were the counterparts of the 'up' workings mentioned earlier. It was also possible to go to Manchester on the local train which gave an extra four trains, though with these it was necessary to change at Sheffield and occasionally Nottingham. Local trains amounted to roughly an hourly service calling at all stations between Leicester and Nottingham. As well as these, certain cross country workings made brief stops at Loughborough on their journey north or south. Loughborough Central in fact was one of the few stations which remained open until the end of British Railways ownership and is the only station on the whole of the London Extension which never really closed, as work was started on renovation by the preservation group almost as soon as the last train ran to Nottingham under British Rail control. The history of this group and the story so far with the preservation of the old Great Central is discussed in more detail in the second part of this book.

After leaving Loughborough Central the line passed by a shunting neck and spur which led off the 'up' line to a couple of sidings which were controlled by the south signal cabin, the north signal cabin being mainly responsible for the station and goods yard area. Between Loughborough and Quorn & Woodhouse the line passed through the Charnwood Forest and the area which is famous for the Quorn hunt. Before Quorn station the village of Woodthorpe could be seen in the distance, one of the communities which the station of Quorn and Woodhouse was supposed to serve along with the two villages which gave the station its name. The village of Quorn lay on the other side of the line and just before the cutting which led up to the platform were the gardens and buildings of the Manor House Hotel which still operates today. Quorn was and still is typical of the Great Central country station. There was a road overbridge from the middle of which a staircase ran down to the two platform faces which the tracks curved round. The facilities at these stations were limited to a ticket office, urinals, and waiting rooms. Only the staircase and the lead from its end to the ticket office were covered by a canopy and this was probably for the benefit of the staff.

In its heyday the staff would have consisted of station master, booking clerk, two porters, a junior porter, two signalmen, a ganger and four platelayers who were all typical Great Central employees in the way they carried out their duties. On Sundays the only staff on hand at the station were the porters who looked after the few trains which passed. In the early days about 300 passengers used the station which would be comprised of clerks, workmen, local landowners and many other wealthy local people. There was often a throng of visitors on bank holidays who came to see such beauty spots as the Charnwood Forest and the quaint little village of Woodthorpe. After they had spent a good day out they were reputed to 'empty the slot machines on the platform and then dangle their legs over the edges and when the train arrived they squashed each other in'.

Other visitors included Royalty, particularly King Edward VIII who would arrange for a London express to be halted at Quorn to pick him up after a day's hunting with the Quorn hunt which frequently made use of the station's facilities. In fact the King even used the ticket office as a changing room while the clerk stood outside. The Quorn hunt, of which King Edward VIII was so fond, often kept the goods yard busy stabling horse boxes which came up for the occasion either coupled to or in conjunction with hunt specials from London. The Great Central produced many special maps showing where the major hunting areas were in relation to the London Extension.

The goods yard as a whole occupied a vast area though it had a very small holding capacity as far as wagons were concerned. There was a store building and a cattle dock each served by a siding, two shunting necks and a connection from the yard to both 'up' and 'down' main line and all these came under the control of the signal cabin which lay in the same position as the new one will, when completed. Unlike most other companies the Great Central station staff were usually responsible for the running of the goods yard and this included handling all kinds of traffic from household commodities to livestock. The yard was occupied by two or three coal merchants who used to collect coal in road vehicles from the railway for delivery and there was also a builder who had his depot here. Goods were delivered on a day to day basis via a train that came from Leicester which would unload its merchandise before picking up full wagons and returning to Leicester.

A fair amount of traffic was made up of food and drink which was for Beaumanor estate, cattle cake for farmers and fertilizers then there was sugar beat for the sugar corporation at Colwick as well as sheep and cattle and loads of timber which again came from the Beaumanor Estate. This timber blocked off the station yard at one point and a crane with a gang of extra men and some flat wagons were needed before the wood could be shifted to its final destination.

During 1944 Quorn was used extensively for troop movements for the D Day landings after which time operations returned to normal. Before we leave Quorn & Woodhouse one other feature is worth mentioning and this is the lettering in the embankment which one can see today proclaiming 'Quorn & Woodhouse'. Originally the lettering was much larger and also displayed the letters LNER and this formed part of a garden in the bank which had been landscaped for the Best Kept Station Competition of 1927, sadly it won no prize and only merited a small mention in the 'Leicester Mercury'.

South of Quorn & Woodhouse the railway passes through yet another area of outstanding scenic beauty, through Buddon Wood to the left of which lie the Mountsorrel Granite Quarries, while the Charnwood Forest and Beacon Hill which rises some 248 feet above sea level are very much in evidence. Slightly further along the line becomes viaduct borne to cross the Swithland Reservoir from which the best view of the area can still be obtained. Although not an important consideration as far as passengers were concerned Swithland Sidings which would have been designated a goods yard had the station been built, were to play a major part in supplying the Great Central with traffic from the granite quarries. The area contained four ordinary sidings, four sorting roads, a cattle dock and a shunting neck, a signal and shunter's cabin and store. The four sorting sidings mentioned earlier were connected up to the Mountsorrel Railway which was a small system that connected up with the Midland Railway at its other extremity.

Swinging away to the right the line reaches Rothley after a fairly short descent which is served by an island platform and goods yard similar to that at Quorn & Woodhouse. Again Rothley served only a small population in true Great Central fashion though it was a marginally larger one than Quorn. One slight difference in layout was that there was a relief siding on the 'down' line side in front of the signal cabin and there were also slightly more sidings in the main yard. This may have had something to do with the fact that Rothley was close to the busy Mountsorrel and Swithland complex as well as Belgrave & Birstall which was the next station in the direction of London.

Belgrave & Birstall, situated on the very outskirts of the City of Leicester was bounded by houses on both sides and was located in a cutting which did not leave any available room for accommodating freight traffic. This station was designed to serve outer suburban requirements and also those who felt like travelling up to the local golf course for a round as there was a rather handy course. Belgrave & Birstall was built in the same style as Quorn & Woodhouse and Rothley except for the fact that there was an extra building provided on the platform. To gain access to the station the Great Central had to build their own road from the main Leicester to Loughborough Road and this new stretch was also beneficial to the local golf club.

Along the line from Belgrave & Birstall the railway began to enter Leicester properly and the first feature which the Great Central had incorporated as part of the large complex was a set of coal sidings at a place called Abbey Lane which had road access and were provided with their own signal cabin to control operations. Having negotiated several obstacles which were crossed by viaducts the Great Central station at Leicester was reached. The first evidence of its approach was a signal cabin and Leicester North cabin. The station itself was a marvellous achievement constructionally speaking, like Nottingham Victoria as so many tracks and a large concourse were fitted into so small a space. Central was not the smallest station in Leicester – this honour fell to West Bridge. Other stations included the Midland to London and Derby, and Belgrave Road which served Lincolnshire and Market Harborough.

The buildings at Central were impressive and like Nottingham Victoria and the other London Extension stations were built in Jacobean style. The frontage incorporated three large entrances flanked by windows, a parapet with nine beautifully formed gables which was later replaced by British Railways with a grotesque plain brick affair. These nine gables had a bulky clock tower which looked disproportionate and on top of this was a cupola in the shape of an onion. The finish was in terra cotta with orange colour bricks making it pleasing to the eye. Above the entrance to the parcels depot was a fine inset inscribed 'Great Central Parcels Office' which remains to this day. Leading in from the street the walkway to the ticket office and platforms was covered with a glass canopy. From the ticket office and booking hall, which was quite a spacious bright area, a rather gloomy and quite unpleasant subway by contrast, led to the trains, though once one had passed through here and walked up on to the platform there was a marked difference. For the trains there were four bay platforms which were mainly used by the local workings and in latter years for the storage of parcels vans. Either side of the tracks which served the outer faces of the platforms, ie the

A busy scene at Leicester Central shortly after the turn of the Century with the ticket examiner setting the departure times and destinations for trains to the north. The 'all stations to Nottingham' train is due to depart from the 'down' bay at 3.55pm while the time is not yet fixed for the departure of the Bradford train from the 'down' main line platform. In the box the ticket examiner is standing on are the various other destinations attainable from Leicester Central by cross-country trains and main line expresses. (Leicester Museums)

main lines, there were passing loops for goods trains and storage of carriages and engines on occasion.

Turning to the platforms in the centre were the respective buildings that one normally associates with a station this size and these were all covered with a large glass canopy and were styled in the same way as the frontage on Great Central Street. As well as the waiting rooms and lavatories for the third and first class passengers there was a dining room and refreshment room which was internally decorated with woodwork and tiles bearing a wheel motif. The restaurant lasted until 29th December 1951 and when it did close it still had many items of crockery, plates, cutlery and the old fittings from Great Central days proper.

On the booking office side of the platform there was a small yard for parcels traffic and a turntable for turning locomotives. Leicester was often used as an exchange point for trains to the north which could pick up a fresh engine for the rest of the trip from Leicester shed which is described in the next chapter. The shed and the main goods yard were situated further south from the station where there was also sufficient room to accommodate the carriage shed, wagon repair shop and all the other facilities which have been more fully described in the previous chapter. The goods yard at Leicester in fact handled all kinds of commodities, from minerals to fabrics, engineering items, footwear and other such industries that were associated with Leicester. The coming of the Great Central London Extension, as with the other lines which had established a presence in the city, made it that bit more prosperous as here was yet another mode of trans-

An inspector poses outside his office at Leicester Central area 1902. (G.C. Museum, Loughborough)

port for bringing in goods and despatching them.

After the goods yard at Leicester the scene soon changed quite dramatically from an industrial and built up residential area to open country as trains shot along Aylestone fields and over the Leicester to Birmingham railway towards the next station which was situated at Whetstone. Today the length of line between Aylestone and Whetstone station has been turned into a public walkway and it is almost hard to imagine that trains once ran along its course until of course one reaches the viaduct just to the north of Whetstone station. The station was built on the island platform principal and access was gained via a staircase which led to the platform which was covered by a rather decorative glass canopy that was incorporated in the main canopy that led to the ticket office building. Whetstone

Leicester Central with 'Black 5' No.44835 simmering in the 'down' platform, 15th July 1966 as it waits for some custom to fill its small complement of coaches, before leaving for Nottingham Victoria. To brighten up the appearance of the engines which were in nothing short of abysmal condition and often breaking down, the drivers chalked names on the side of the smokebox and smokebox doors such as 'Wild Thing' a reference to a pop song of the day, which probably also had some bearing on how the engine ran! (K.C.H. Fairey)

Whetstone station prior to full completion showing all the platform buildings. The glass canopied section on the left covered the stairs up to the concourse from the road below. The first building on the platform is the booking office at which a solitary passenger is purchasing a ticket. Building number two housed the waiting rooms and the ladies' toilets while the third structure provided the gentlemens' lavatory. ***(Leicester Museums)***

only served a very small community hence the small island platform. The village though did have one large industrial area which lay off the Cosby Road and this complex was part of the General Electric Company, which also had a plant at Rugby that was served by sidings leading off from the Great Central. The plant at Whetstone however had a more tenuous connection with the Great Central in the sense that a locomotive named GT3 was built there and ran tests on the Great Central main line, at a time when British Railways were still evaluating all kinds of traction that might replace steam. The station's chief commodities were coal, stone and livestock which is evinced by the fact there was a stone dock, cattle dock and horse dock.

Just south of Whetstone the line curved round gradually near the village of Cosby and this curve got the nickname the Cosby Corner by drivers as a result. The line from here followed a more or less straight course borne on embankments and through cuttings but here if any, it is possible to see the main reason for the downfall of the Great Central main line, because there were vast open fields on either side of the line and only the odd farm house or group of cottages visible. Mid way between Ashby Magna and Whetstone lay the old Rugby to Leicester railway which meandered through the villages of Ullesthorpe, Broughton Astley, under the Great Central to Wigston where it met at a junction with the Midland main line and the London & North Western to Birmingham. Turning back to Ashby Magna an island platform was installed here for the benefit of the tiny village which it served and in layout this was similar to the one at Whetstone apart from the fact that the access was made to the platform from the road above. The village itself was very small and had no industry apart from farming and with the coming of the railway there was very little change, apart from a few inhabitants who

A few customers wait at Ashby Magna for the Rugby train. All facilities on the station were boarded up by this time including the ticket office as tickets were sold by the guards on the train to reduce operating costs. Closure is imminent judging by the three notices and their proposals, on the billboard. *(N.G. Steele)*

decided the conditions were better on the railway and so ended up working on the station.

Ashby Magna had its own signal cabin as had every station along the line and a small goods yard which catered for domestic goods such as coal and the like. The goods yard was later eaten away partially with the construction of Britain's premier motorway which was another contributor to the downfall of the London Extension. Ashby Magna was surprisingly, one of the few stations which survived to the bitter end being retained during the semi fast period of operation where trains stopped at between seven and nine stations on their journey from Marylebone to Nottingham or vice versa. It can only be assumed that the stop was put in to help make the journey as long as possible in an attempt to kill the line.

Another station which was retained until the last day of British Railways operations was that at Lutterworth which was situated on one of the high speed parts of the whole line. It is perhaps not difficult to envisage why this station remained open as it served a larger community altogether which boasted a fair amount of industry in the last 30 or so years. Lutterworth had some 5,000 inhabitants and its proximity to Rugby, which was a major centre, made it an attractive area in which to live. The station lay to the east of Lutterworth and was again on the island platform basis and it played host to local trains and occasionally some express workings in the early years, though by 1960 it became a regular port of call until 1966. It remained for the shortened Rugby to Nottingham commuter service until 3rd May 1969. Unlike other stations one of the buildings on the platform was isolated by the station entrance which was effected from the road below.

There was a stationmaster's house which still stands today and the traditional fir trees which the Great Central planted to screen the railway. The signalbox is worthy of note mainly because it was a rather large structure for such a small station, but of course it had been intended to add extra lines and enlarge the goods yard as was clear from the extra space available on the lever frame. It was even

Lutterworth station, the last stop before Rugby which marked the end of the northern section of the Great Central. On the left is the station master's house while the platform buildings and access bridge can be seen in the background. Judging by the height of the trees in front of the station the photograph was taken shortly after the opening of the line. The pine trees were a feature of the London Extension designed to hide the imposing structures of the railway and in time Lutterworth became shielded by a blanket of trees. *(Leicester Museums)*

evident from the size of the land owned by the Great Central that some sort of extension was planned. This was probably regarded as a possible overspill area for Rugby.

It was just south of Lutterworth in later years, at Shawell where the motorway parted company with the Great Central after following it for some 19 miles. From Shawell the line arrived on the outskirts of Rugby which was the end of what was regarded as the first section of the London Extension. This was marked by a massive viaduct which crossed the valley through which the River Avon flowed. Towards the end of the viaduct the main London & North Western 14 track line was bridged by a large girder structure which carried the Great Central tracks towards the latter company's station. This large girder section gained the nickname the 'Birdcage' because it was akin to a cage-like structure. Shortly before it was completed the Great Central was made to erect an additional layer of signals on one of the LNWR's gantries as the original signals could no longer be seen from the signal box which controlled them. This mass of signals on such a large structure also had a nickname, the 'Bedstead'.

Although Rugby station only had one platform built on the island basis it had to serve a large centre so it was graced with more buildings than usual and was similar to both Brackley Central and Loughborough Central. Like the latter two the booking office was at street level and was on the Hillmorton Road. A staircase led down to the main concourse which contained three waiting rooms and a toilet block which was the only building not encompassed by the overall canopy.

Rugby Central station was one of the stops for main line trains and also the start of local services to Nottingham, Leicester and Aylesbury. Later in life, between 1966 and 1969, Rugby became the starting point for the truncated commuter service after the line had closed as a through route. With its population of 57,000 Rugby definitely had need of two stations and with the town's engineering and other heavy industry there was much need for a second railway. In fact the Great Central, as mentioned earlier, afforded a connection to the General Electric Company factory. This led off from the yard which was slightly larger than most, being provided with more sidings, a goods warehouse and various offices. The engine shed and fitting shop lay at the end of the sidings which served the horse dock. The whole area is now used as a timber yard

Right, top: Rugby Central viewed from the London side of the platform was similar in design to the platforms at Brackley, Woodford & Hinton and Loughborough, which were provided with more lavish facilities than the ordinary island platforms of the London Extension because they served larger centres. Notice the strange design and higgledy-piggledy layout of the buildings which was mainly due to the fact that the road bridge did not run at right angles to the platform. This made construction of the booking office and staircase rather awkward. *(Leicester Museums)*

Right, bottom: Rugby Central looking towards Nottingham in the days of the semi fast services. *(G. C. Museum, Loughborough)*

RUGBY
CENTRAL

by a merchant who expanded his business after the railway closed having already established a presence there.

The next station along from Rugby Central was Braunston & Willoughby which was situated along the Daventry Road and like many Great Central wayside stations, it was built with economy in mind having the most spartan facilities. In effect both Braunston and Willoughby villages were miles from the station and this was why the station, along with Culworth became an early casualty as far as closure was concerned. Originally the station had been just called Willoughby and in a desperate attempt to try to attract more trade the name Braunston was added. A similar trick was adopted at Finmere which lay on the Buckingham Road and which at about the same date became known as Finmere for Buckingham. Buckingham though was served by a station which actually lay quite close to the town centre. Braunston had already been catered for by the London & North Western on the line from Daventry to Leamington. A little further along the line between Willoughby, and Staverton and Catesby viaducts lay Staverton Road signalbox which was basically a small wooden cabin. This was oil lit and isolated from all civilisation.

Between the north end of Catesby Tunnel and Braunston & Willoughby the famed 'Windcutters' were reputed to use the stretch for attaining speeds well into the eighties and nineties. Catesby Tunnel was constructed purely at the wishes of the occupants of Catesby House who were descendants of the notorious Robert Catesby, who was executed for his part in the Gunpowder Plot. This was due to the fact that they did not want to see the railway, which would otherwise have passed through a large cutting. The tunnel was some 3,000 yards long and proved very difficult to build, especially with the constraints put on where ventilation shafts could be sited, so as not to offend the local residents. The thousands of bricks which lined it always dripped with water and soot, often into locomotive cab roofs. The atmosphere has changed very little today.

At the south end of the tunnel where the track emerged, the line swept through a cutting and round a gentle curve into Charwelton station. This was a busy little exchange station for ironstone traffic for a local quarry which had its own sidings leading out of the main yard. Access to the station was made by a stairway from the main Daventry to Banbury

A vintage scene at Charwelton station with one of 'the Railway Children' standing between two smartly dressed members of GCR staff. ***(Leicester Museums)***

Cause for celebration at Charwelton as the bunting is out and a Union Jack flying from the porter's room chimney. Part of the platform has been cordonned off on the London side for the Royal Train which can just be seen in the background. Note the station nameboard, underneath which are a series of interesting timetables and posters. On the top row are a comprehensive list of train times covering the whole of the Great Central network while on the bottom a number of different excursions organised by the Great Central are on display covering all areas at home and abroad. *(Leicester Museums)*

road, though the bridge has now disappeared with modern road improvements. The station itself was provided with the usual facilities that were associated with an island platform that served a rural community. There was also a stationmaster's house in the yard, similar to those provided at most of the other stations. Returning to the yard, it was always busy and invariably there were anything up to 220 wagons stabled at a time. An average day would see 18 trains pass the signalbox in addition to passenger trains which roared through or stopped at the tiny station. The busiest time for Charwelton station and yard was probably during the Charwelton Week when much livestock was handled in the goods yard. Occasionally the odd lost sheep or cow would provide a source of humour as it was found wandering away down the line. On leaving Charwelton the line ran past a set of water troughs which were fed under pressure from their own tank. These allowed engines to replenish their water stocks without stopping, speed being of the essence on the Great Central as per its main slogan, 'Rapid Travel in Luxury'.

After a short run from the troughs Woodford & Hinton was reached which was probably one of the most interesting places in the whole of the Railway's history. Until the arrival of the Great Central it had been nothing but a sleepy little village in the middle of the Northamptonshire countryside whose nearest railway was the East & West Junction Railway. It was not without good reason that it was always regarded as a somewhat 'Erratic and Wandering Journey Round'. Two links were forged with the

Great Central from this line, one which enabled through trains to work to Stratford upon Avon but later this was closed and became carriage sidings.

The staff at Woodford & Hinton station take a break during a slack period to pose for the photographer. Note the variations in uniform especially where the hats are concerned. On the right the platform for Stratford and banbury trains can be seen.
(Leicester Museums)

The other link started at Woodford where the branch had its own wooden platform for local trains to Byfield or even to Stratford upon Avon. This wooden platform later became used by trains working to Banbury from Woodford until they ceased to run in 1964. These were known as the Banbury Railmotors. In 1956 the platform was modernised when replaced by a concrete structure. Alterations were also made to the footbridge which led to the main island platform. This was something of an anomaly as far as design was concerned as it contained more buildings than those stations at Brackley, Rugby and Loughborough which served large centres. There was even a refreshment room which survived until 1954 when it eventually succumbed due to lack of custom. The lavish station was provided for the benefit of the large railway community as well as passengers who wanted to set out from this 'little Swindon' miles from anywhere. The station boasted three additional blocks of buildings in addition to the refreshment room, having all the usual facilities such as toilets, cycle store and store rooms, waiting rooms for first and third class and a porter's room as well as a ticket office. This was unusual as it was not situated at road level as at most other large stations, and was probably due to the fact that Woodford had its access from a staircase built into the bridge which carried the tracks over the road, and this was quite narrow.

Opposite the station on the east side was a small goods yard which had an access road leading up from the main thoroughfare through Woodford Halse village. This yard had two sidings, one of which led to a warehouse while the other was situated near the engineer's workshops. These and the loops round the main platform tracks and the junction plus the numerous turnouts in the station area, were under the control of Woodford No.4 box. North of the station, between the main tracks and loops, lay a locomotive holding siding. Leading off the main lines on the west side there was a large yard on an embankment built using spoil from Catesby Tunnel and contained 16 sidings, while the one on the east side had 15 sidings. These yards came under the control of No.3 signal box. Signal boxes Nos 1 and 2 were responsible for the control of new yards which were added in 1941 to cope with extra traffic during the war. Built to the north of the original yards these had 16 lines on the west and 12 on the east of the main line. These new sidings were to prove a great benefit to traffic on the London Extension in 1947 with the introduction of a fast freight service running between Annesley in the North and Woodford in the South. Comprising 40 trains which ran with precision timing and known as 'Windcutters' and 'Runners' they were an unbeatable service and well worthy of their names which were adopted when the 9F locomotives were draughted in to work them.

Woodford was always noted for its intense freight workings and in the early years of the Great Central, drivers carried fish, coal and steel, mainly on lodging turns and it was not unknown for locomotive crews to be given the odd perk for their endeavours. The fish came from Grimsby and Immingham while the steel traffic was from Sheffield as a result of the manufacture of cutlery and household utensils. The coal was needed for internal as well as commercial use as it not only provided fuel for the locomotives but also a much sought after means of heating for homes and energy for industry. Coal traffic ensured a staple revenue for the Great Central. Woodford also had a role as a wagon exchange point for the other major companies in the area, such as the Great Western and the East & West Junction Railway whose complex connecting lines south of the station, made a variety of destinations possible from the massive sorting and marshalling yards at Woodford Halse depot. To accommodate these trains' engines a massive engine shed existed between the old and new east yards which had capacity for 30 locomotives as well as a whole host of facilities which are described in the section on locomotive sheds. Near the depot there was a separate set of wagon and carriage shops which could carry out minor repairs. The wagon shop was served by a traverser which was on an adjacent siding.

Before leaving Woodford one other feature which is worthy of mention was the social club which was formerly a country hostelry known as the Hinton Gorse or Goss as the locals called it. The building, which was a prominent landmark in the village, had always had some dealings with the railway and huntsmen used to frequent it having come up from London on the Great Central hunt specials. In 1940 perhaps not surprisingly, the railway acquired the premises for the benefit of the many railwaymen who lived in the community as there was no other form of entertainment save for a small cinema. By 1955 the building was taken under the wing of the British Rail Staff Association and many social events were staged there though the club became largely redundant once the depot closed ten years later. However the enthusiasm of some employees who had worked the line was such that it was decided to form an Association, which still holds meetings where one can relive scenes of the Great Central and other railways in steam days. Meetings are held every second Wednesday of each month throughout the year.

Steaming away from Woodford's large complex and nostalgia, the line passed a small isolated signal cabin which served the junction with the Banbury line, before approaching Culworth, which did not really serve anywhere! The Great Central's Culworth station was nearer the village of Moreton Pinkney than Culworth which was actually better served by Eydon Road halt on the Banbury line. Moreton Pinkney anyway, had its own station already on the Northampton to Stratford line. It was Culworth's location that was to make it an early casualty as far as closure was concerned – it being closed to passengers in 1958.

Culworth station several years after closure but in a surprisingly good state of repair considering its long period of abandonment. The station was closed on 29th September 1958 because, like its counterparts which ceased to be passenger stations on 4th March 1963, it served an area of scant population.

(G. C. Museum, Loughborough)

Along from Culworth the line ran through a series of cuttings and embankments before reaching the splendid nine arch viaduct over the River Tove and the London & North Western Railway, which

HELMDON
FOR SULGRAVE

still stands today. After the viaduct the line reached the little island platform at Helmdon which was served by local trains between Aylesbury and Rugby. Helmdon like the other local stations had its own goods yard which was mainly used for the delivery of coal and other domestic necessities. The station lost its passenger service on the 4th March 1963 with the removal of the local service north of Aylesbury, while goods traffic ceased on 2nd November 1964. Facilities on the LNWR branch

Left: The beautifully kept island platform at Helmdon on the day of closure to passenger services, 4th March 1963. On the station nameboard the words 'For Sulgrave' were included to encourage any would-be visitors of the ancestral home of the Washington family to travel via the Great Central station. When this picture was taken the 12.38pm ex Marylebone was due for Nottingham Victoria. Just to the north of the 'down' home starter signal was Helmdon viaduct which crossed the River Tove and the old Northampton and Banbury junction line.
(G. C. Museum, Loughborough)

station had been taken off some 13 years earlier in 1951 when the line from Cockley Blake to Greens Norton Junction was closed.

Three miles on from Helmdon, after a continuous deep cutting, Brackley Central was different to the other island platforms in two respects. Namely it was the last major station on the London Extension before Marylebone, while the other difference was that it had a side access rather than the normal central entrances like those at Loughborough and Rugby which were otherwise built to similar specifications. There were fears that the station would cause terrible traffic congestion on the main Towcester Road. A lay-by was provided in front of the booking office for carriages, and both features remain today. Further down the road there

Left: Brackley Central was one of the larger island platforms on the London Extension as the town had quite a sizeable population. In this picture the most interesting point as far as the construction was concerned can be seen. Originally the platform was supposed to have ended at the bridge in the background so that a staircase could have been built from the road to the platform giving access to passengers. However, there was concern that this access might cause a bottleneck on the main road from Oxford to Towcester so the Great Central was forced to build a lay-by further down the road. This gave access to the booking hall and general offices that were built into the bank on the left.
(G. C. Museum, Loughborough)

was another entrance leading into the goods yard which boasted a large shed handling the staple commodities of the town. Down the other end of the town there lay the London & North Western station which was on the Verney Junction to Banbury Merton Street line which closed in 1964. Brackley Central however remained open until the final closure for passenger trains and was occasionally used as a stop for express workings in its heyday.

The line left Brackley Central on a splendid 23 arch viaduct that spanned the Ouse Valley, from where it ran until it reached the penultimate station at Finmere. This was another lonely island platform with a small goods yard with its own cattle dock, coal staithes and store. Finmere was served by local trains and its busiest times were always at the beginning and end of term time at Stowe School which lay just across the fields. An additional way of serving the station was inaugurated in 1923 by means of a slip coach which dropped off the back of an express; this process being continued well into London & North Eastern Railway days. After being dropped the coach would be worked forward to Helmdon and then taken empty to Woodford. Undoubtedly a contributing factor to the slip working was due to the presence of a number of distinguished people living in the locality who would have welcomed such special treatment.

The last port of call on the London Extension before it joined with the Metropolitan, whose lines the Great Central shared from a point just north of Quainton to London, was Calvert which lay to the south of the main Oxford to Cambridge line. Calvert had originally consisted of one or two houses and had no name as the place was not large enough to merit one. The derivation of the name is indeed interesting in itself as it stems from the Verney family who were local landowners. Sir Harry Verney, who was a very keen railwayman had been born a Calvert by marriage and after he had inherited the family estate at Claydon he became a Verney. His growing liking for railways prompted him to dedicate a station to his mother when the Great Central arrived. This was called Calvert, around which area in later years a brick industry and small community arose. As well as bricks the yard at Calvert dealt with coal, timber, milk etc. The station platform still exists today and the 'down' main line is used as a siding for the Greater London Council waste trains, while the 'up' line provides a link from Aylesbury to the Bletchley line along which diesel multiple units from Marylebone travel for servicing.

The station at Marylebone was a modest affair though very well appointed, with a refreshment area in the form of two bars and a dining room. There was a beautiful oak panelled ticket office with four ticket windows facing out into the booking hall which had access from the street where there was a pull up for taxis and carriages. A large awning led to a building across the road and this was to keep travellers dry on the short trip to the Great Central Hotel which was once a very grand structure with a well appointed interior, and boasting its own cycle

An early postcard view of Marylebone station showing the main concourse with a reasonable number of passengers in evidence. The booking office is visible to the left with the refreshment and dining rooms in the block behind, while in the middle stands the ornamental garden. On the right by the entrance to the platforms is the W. H. Smith's news stand along with a couple of others which have long since gone. *(G. C. Museum, Loughborough)*

The impressive and prominent 222 Marylebone Road in its latter day role as offices for the British Railways Board. After 87 years of railway use, the building has been sold and an alternative use is to be found. During the early years of the London Extension '222' served as an hotel for weary travellers from the north. With the absorbing of the Great Central into the LNER, the hotel was turned into offices. *(G. C. Museum, Loughborough)*

track which ran around the roof! This building now known as 222 Marylebone Road still exists, used until recently by the British Railways Board as its Headquarters. Though sadly, most of the finery has been removed from the interior. Both the station and the former hotel make a pleasant sight from the exterior even today to the thousands of commuters who use Marylebone.

As well as the ticket area and refreshment rooms, the station also had a set of ladies' and gents' toilets and waiting rooms, an enquiry office, a first aid room, a parcels office, a telegraph room, station master's office and medical officer's room, together with a number of stalls which were positioned on the concourse as well as an opening which led down to the Bakerloo Line. The whole layout was very pleasant and even the offices on the upper floors were designed so that as much light as possible would be able to penetrate the canopy which covered the whole structure, and then work its way round the sumptuously appointed corridors of the Great Central management.

Turning to the platforms there were only four built which were used for both express and local workings, but there had been a plan at one time to expand the station when the need arose which would have meant engulfing part of the huge goods yard which has now been demolished. The first structure to be demolished in the goods yard was the magnificent warehouse which was hit by a bomb during the Second World War. The goods yard lasted in service until 1961 when the land was sold for housing developments though it had always been busy and responsible for the shifting of heavy goods as well as domestic goods as was evident from the provision of a large electric crane. At one time the yard was used to put on a display of Modern Traction of British Railways, which included locomotives such as GT3, an a.c. electric locomotive, a 'Warship', a 'Deltic' and one or two other classes of diesel locomotive. On the opposite side to the goods depot across the main lines there was a large carriage shed and washing facilities together with a long milk dock which served a dairy depot.

Marylebone station from the exterior on 5th April 1966 a few months prior to it ceasing to be an important mainline terminus. Unlike St. Pancras, King's Cross, Paddington and Euston, Marylebone was plain in appearance and included none of the intricate architectural features designed into the other structures in London. The main reason for the lack of ornamentation around Marylebone station was due to the fact that the Great Central did not have sufficient capital left after the building of the London Extension. In the middle is the covered walkway leading from the booking office to the old Great Central Hotel. Note also the gates in the foreground, a set of which were provided at each end although sadly these have since been removed.
(G. C. Museum, Loughborough)

Chapter 3
On the Line

Express and Through Services

In addition to the lavish facilities provided at the various stations along the route, the Great Central lived up to its promotional jingle 'Rapid Travel in Luxury', by providing an excellent train service that was considered second to none by those who used it. As far as special or named trains went the Great Central believed that the services they ran equalled the quality of any thing provided by Pullman Cars which were used on other routes and like the Great Western saw no real need for dressing trains up or giving them fancy titles in order to attract patronage. This remained the situation until the end of the London Extension route, with ordinary carriage stock being used for all trains. However, 'The Master Cutler', which is mentioned later in this section became a British Rail Pullman service when it was transferred to King's Cross from Marylebone where it worked Sheffield services for a short while. Later it was moved again to work out of St Pancras station.

The origin of 'The Master Cutler' service is a very interesting one and dates from the early years of the Great Central when the company ran a train dubbed by the press as the 'Sheffield Special'. Not

'Director' class 4-4-0 No.5505 **Ypres** *waits at Marylebone station in 1924 with a 'Sheffield Special' train comprised of Great Central 60ft teak stock. No.5505 was claimed to be the best performer of its class and on one occasion it worked the Manchester turn for four weeks without a break.*

(Ken Nunn/LCGB)

only did it have special significance in the line's history but was also a crack train. Originally the train was destined to be a Manchester working but the Great Central felt that it ought to try to tap the South Yorkshire market by running to Sheffield. This was a bid to run as far as Leeds Central station where it would be able to provide a much needed service with rapid travel to London for millowners and other travellers. The Great Central believed that it was within its rights to exploit the potential, but the competition did not agree and sought an injunction to prevent running into Marylebone direct from Leeds. The intention had been to put on a train that would leave Leeds at 7.20am calling at Wakefield, Rotherham and Sheffield before making a non stop run to Marylebone. By 1903 the chance was finally given to the Great Central with the rescinding of the court injunction preventing the running of trains.

In order to obtain the maximum potential from the Sheffield service a slip coach for passengers wishing to alight at Nottingham or Leicester was attached. There was a corresponding service from London Marylebone which left at 3.25pm taking the same route. Later, because of the intense working competition of the Great Northern the Great Central was forced to run the train to Manchester via Sheffield, where a change of train was now necessary for Leeds. Still concerned about the lost Leeds service the Great Central tried a new ploy, namely to slip a coach at Leicester which was taken on to Cleethorpes. The latter coach had the advantage of being attached to a train which stopped at a number of places before reaching its final destination. Over the years a number of changes took place in the operation of the 'Sheffield Special' and eventually it was discontinued prior to the Second World War.

Another interesting train was the Immingham boat train which ran in connection with Orient Line cruises to the Norwegian fjords. This service ran from Marylebone along the London Extension carrying full restaurant facilities and had the distinction of running onto the quayside to be alongside the ship. In later years a similar train was the 'Port to Ports', the Newcastle-Swansea service calling at Darlington, York, Sheffield, Nottingham, Leicester, Banbury, Cheltenham, Gloucester, Newport and Cardiff. Other through services which

Atlantic No.1086 draws away from Whetstone with a fine head of steam making its way on towards Ashby Magna. The train, a Leicester to Marylebone express is comprised of an interesting variety of rolling stock with a Parker 50ft brake third leading while the second is a compartment brake third. The third vehicle is a restaurant car followed by two more 60ft vehicles leaving the last two carriages, a clerestory and a 6-wheel brake van.
(V. R. Webster)

B17 class No.2860 **Hull City** *with a Marylebone to Leicester stopping train passing the lonely gas lit signal cabin at Staverton Road in 1946. (J. G. Coltas)*

Double headed by 8B class Atlantic No.263 and 11D class 4–4–0 No.1020, a pair of 'Barnum' saloons and eight 60ft carriages form a 'down' Manchester train near Leicester. (V. R. Webster)

A 'down' Bradford train headed by C4 class 4–4–2s Nos.6090 and 5263 near Rugby on the 2nd April 1929. The first three coaches are a 60ft compartment carriage, followed by a Parker vehicle, both of which are minus their headboards. The third vehicle is a further 60ft type while the rest are Gresley stock.

(W. L. Good)

Leicester based A3 4–6–2 No.60052 **Prince Palatine** *heads out of Marylebone's platform 2 with a rake of Gresley coaches newly painted in the British Railways' blood and custard livery. This was 'The Master Cutler' service to Sheffield, shortly after its inauguration.*

(G.C. Museum, Loughborough)

THE SOUTH YORKSHIREMAN
61177
61177

Left: Rebuilt 'Royal Scot' 4–6–0 No.46107 **Argyll and Sutherland Highlander** *on a Nottingham – London semi-fast at Evelyn Road Leicester, 11th January 1962.*

(John Clarke Album)

operated all the year round included one from Manchester to Bristol which called at Huddersfield, Halifax, Sheffield, Nottingham, Leicester, Banbury, Oxford and Bath. There was a Manchester, York or Bradford service to Bournemouth via Oxford and Southampton and York to Southampton runs. Shorter runs such as Woodford to Southampton also existed as well as Nottingham to Skegness, Leicester to Sutton-on-Sea, though these trains were mainly summer workings only. As far as the latter were concerned if it had not have been for the link between the Great Central at Woodford with the Great Western at Banbury there would have been a very restricted potential for Great Central through workings.

Left, below: B1 class 4–6–0 No.61177 heads out of Nottingham Victoria with a train of Thompson coaches forming 'The South Yorkshireman' service to Bradford in April 1956.

(G. C. Museum, Loughborough)

Below: B1 class No.61185 leaving Leicester Central on a relief Sheffield to Bournemouth train, the 6th August 1955.

(J. F. Henton)

Through workings continued using the London Extension until the final years of its existance although the line saw more than its fair share of internal workings in both the passenger and goods sectors. In the main, the pattern of services was made up by dividing trains between four destinations. In the early years these were Sheffield, Manchester and Leicester and in the last six years, Nottingham which acted as the terminus for the drastically pruned semi-fast services. After 1960 only three trains a day ran each way between Marylebone and Nottingham carrying as few as three coaches on occasion and these left Marylebone at 8.40am, 12.40pm and 4.30pm and from Nottingham they left at 8.40am, 12.25pm and 5.15pm. The turning point in the fine career of the Great Central as a major trunk route occurred in 1960. Prior to then a diagram of seven trains had operated each way for most of the line's life, beginning with a newspaper train from Marylebone which was then followed by three expresses to Manchester London Road leaving at 10am, 12.15pm and 3.20pm. The latter was the old 'Sheffield Special' train. The next

train of the day was 'The South Yorkshireman' which, like 'The Master Cutler', was introduced just after the Second World War. It is believed that this 4.50pm departure was an attempt to try to succeed where the 'Sheffield Special' had failed in previous years.

'The South Yorkshireman' called at Aylesbury (running via High Wycombe) Rugby, Leicester, Nottingham, Sheffield, Pennistone, Huddersfield, Brighouse, and Halifax before terminating at Bradford. The penultimate departure of the day was that of 'The Master Cutler' which was the other crack train and this left at 6.18pm, travelling as far as Sheffield calling only at Rugby, Leicester and Nottingham en route. The last service of the day was that of the 10pm mail train for Manchester and Liverpool Central.

Services into Marylebone started with 'The Master Cutler' in later years, leaving Sheffield at 7.40am, followed by 'The South Yorkshireman' from Bradford at 10.00am. Then it was the 2.10, the 1.05 and the 6.15pm from Manchester, finishing off with the return mail service from Liverpool. For the most part trains consisted of eight coaches plus a

***Right:** Class V2 2–6–2 No.60820 passing Leicester Goods Junction South with the 12.20pm York – Banbury working on 2nd May 1953.* *(N. G. Steele)*

refreshment car except during the summer months when it was not uncommon to see a locomotive struggling with twelve vehicles. Returning briefly to 'The Master Cutler' it displayed the elitism of the Great Central and was rather akin to the atmosphere of a gentleman's club because it enjoyed a strong patronage and its clientele were generally businessmen from Nottingham or Leicester.

In 1958 under intense competition from other trains running to Sheffield the Great Central route lost its prime service to the new and faster Pullman running out of King's Cross and the 'Robin Hood' which ran to Nottingham from St Pancras. After a long period of indecision the 'Robin Hood' and the Pullman were finally axed and the 'Cutler' to this day, along with 'The South Yorkshireman' now operate out of St Pancras. Even today 'The South Yorkshireman' has not improved its rather doubtful reputation as it was always considered the poorest revenue earner, but this was mainly due to the ludi-

***Right, bottom:** B16 class 4–6–0 No.61420 leaving Leicester Central on a relief train from Scarborough, 6th August 1955.* *(J. F. Henton)*

BR Standard Class 5 No.73066 passing Swithland Sidings Box near Leicester with a return Holiday Special from the south coast, 15th August 1964. ***(M. D. Marston)***

476
61420

45708

Left: A B1 class 4–6–0 No.61175 with empty stock for a Nottingham Victoria to Mablethorpe Special leaving Bagthorpe carriage sidings at 8.03am, Saturday 27th July 1963.
(T. Boustead)

Left bottom: 'Jubilee' class 4–6–0 No.45708 **Resolution** *seen here emerging with its train from Weekday Cross tunnel on a Newcastle to Bournemouth working in the summer of 1962.*
(G.C. Museum, Loughborough)

crous times that it made its runs. Both trains changed engines at Leicester and in the case of the 'Yorkshireman' there was a rather interesting quirk in the operation as it was a Bradford and not a Leicester engine that used to haul the train to its eventual destination. On the way down from Bradford, engines booked to work 'The South Yorkshireman' also looked rather odd because they used to carry the headboard reversed until they reached Leicester where it was turned round for the rest of the journey.

Local Services

Local services on the Great Central did not really develop until the early 1900s mainly because of the desire to attract revenue using the long distance services. Also the Great Central needed to market the product first before it could actually provide the service. Between Annesley and Aylesbury the Great Central had a record number of local stations with very few main stopping points. Apart from Nottingham, Leicester, Rugby and Aylesbury, other stations which enjoyed a main line service were Brackley, Woodford & Hinton, Lutterworth, Loughborough and New Basford. These stations also enjoyed a local service that made inhabitants of the small villages down the centre of England more mobile. At the peak of London Extension operations, the busiest section for local services was that between Nottingham and Leicester with some twelve trains each way every working day. Some just ran between the two cities while others worked another stage down to Rugby or Woodford for example, which enabled the Great Central to get the best use out of the trains. In effect these workings became long distance stopping trains and to give an idea of the diagram they worked, some two stage services are detailed as follows:

Nottingham–Leicester–Woodford and Hinton	Woodford–Brackley–Marylebone	Leicester–Rugby/ Rugby–Woodford
Arkwright Street	Helmdon	Whetstone
Ruddington	Brackley (stop or continue)	Ashby Magna
Rushcliffe Halt		Lutterworth
East Leake	Finmere	Rugby Central (stop and return or continue to Woodford)
Loughborough	Calvert	
Quorn & Woodhouse	Quainton	Braunston & Willoughby
Rothley	all stations to	Charwelton
Belgrave & Birstall	Marylebone	Woodford (stop)
Leicester (stop and return or continue)		
Whetstone		
Ashby Magna		
Lutterworth		
Rugby Central (stop or continue to Woodford)		

In effect the service boiled down to there being trains between Nottingham and Leicester; Nottingham and Woodford; Nottingham, Rugby and Leicester; Rugby, Leicester, Woodford and Marylebone and Leicester, Brackley and Woodford. There were additional workings between Leicester, Nottingham and Sheffield with corresponding trains in the other direction, giving the line a comprehensive service.

Local passenger trains were generally hauled by almost any type of locomotive from goods tanks to newly outshopped engines or retired express passenger locomotives. At one time the Great Central

Above: The 'Annesley Dido' comprised of two former Manchester Sheffield and Lincolnshire Railway 4-wheel coaches seen here in the charge of Class D12 No.439 at Bulwell Common.
(D. J. Montgommery)

Left: Sacre 2–4–0T No.449 at Woodford & Hinton waiting to work the Stratford-on-Avon train over the spur to Byfield and the Stratford-on-Avon & Midland Junction line. The branch platform where the train is standing was also used for Banbury trains.
(S.W.A. Newton/V. R. Webster)

Below: No.1125, 4–4–2T class 9L on a Banbury train near Woodford, comprised of goods vans and passenger vehicles.
(Real Photographs)

'Royal Scot' No.46118 Royal Welch Fusilier *awaits the right away from Banbury with a three-coach local for Woodford on 24th May 1962. On the right is No.6906* Chicheley Hall, *acting as station pilot with a parcels van.* *(K. C. H. Fairey)*

In this picture of Nottingham Victoria two trains wait in the 'down' bay platforms on the northern side. In platform 2 an ex Great Northern J6 class 0–6–0 waits to set off for Derby.
(G. C. Museum, Loughborough)

Ivatt Class 4MT 2–6–0 No.43155 with the 6.25 pm Nottingham Victoria to Derby Friargate service heading tender first between Bagthorpe Junction and New Basford Tuesday 25th August 1964.
(T. Boustead)

saw the use of three steam railcars built by Brush working between Leicester, Rugby and Loughborough but they only survived for a short period. On one occasion a rather grimey and forlorn looking Pacific No.60103, better known as No.4472 *Flying Scotsman*, was seen pulling a local Leicester to Rugby stopping train whilst awaiting overhaul.

Express Freights, 'Windcutters' and 'Runners'

The Great Central throughout its working life saw a variety of commodities pass over its London Extension line. The first was coal which was carried in heavy trains in the early years to help test how the embankments and other major engineering features were settling, prior to the opening of the line as a trunk route.

As the years progressed the Company began to build up various trade from newspapers, parcels and mail to ironstone, steel, limestone and gypsum. To cope with the various items an elaborate set of sidings was provided at each station and there was the usual inclusion of an animal and milk dock area as well as coal staithes. In addition to the sidings at stations there were holding sidings at various points on the line where interchanges were effected or where the main line joined a private, industrial railway (as detailed later in this chapter).

At Calvert, as well as the general freight facilities a little complex of lines ran into Itter's Brickworks and this also connected with the Oxford to Bletchley line. Local stations which were usually served by a pick up goods service handled small quantities of coal, parcels, milk and livestock from neighbouring farms. Some of the local stations were registered as having no goods handling, such as the platforms at Belgrave & Birstall, Chacombe Road,

Right, top: Class 04 2–8–0 No.6221 passes Charwelton with an 'up' mixed freight for Woodford sorting sidings. The train on the left is a rake of goods empties for the ironstone mines which had exchange sidings in Charwelton yard.
(L. Hanson)

Right: A Class 04 from Langwith, No.6537 approaches Charwelton just north of Woodford with a rake of coal empties for the north on 3rd July 1937.
(L. Hanson)

Class 9F 2–10–0 No.92030 on 31st August 1961 seen passing Staverton Road signal cabin with an 'up' freight for Woodford Halse. *(K.C.H. Fairey)*

Class 04 No.5378 with a short pick up freight, working its way along the London Extension past Staverton Road in 1939 en route to Woodford. *(J.G. Coltas)*

Eydon Road Rushcliffe Halt, Carrington and Arkwright Street in Nottingham. Generally these stations relied on neighbouring facilities. Each village or small town along the line had an agent for the Great Central Railway who could arrange travel facilities if you needed anything shipped anywhere.

The next place along from Calvert to have any significance as far as goods was concerned was Brackley which boasted a brick built goods shed, a very rare feature for a town with such a small population. Similar buildings were to be found at Rugby, Loughborough and on a much larger scale at New Basford, Leicester, Nottingham and Marylebone. Woodford Halse although deep in the heart of the Northamptonshire countryside was a different story altogether as it was really the nerve centre for goods operations, along with Annesley. Latterly these two places were to feature largely in the freight system.

From Woodford to Leicester the stations only handled a small amount of traffic because they served basically rural areas. Leicester's goods yard was split into three areas, the parcels being dealt with in the vicinity of the station, while the coal was handled at Abbey Lane Sidings. Leicester South looked after all other goods on a vast complex of sidings and in a large warehouse at Braunstone Lane.

Moving north to Loughborough Central there were a few additional sidings which served a builder's yard and a timber yard. No extra facilities were needed for the limestone quarries in the area as the Great Central had provided a large series of exchange sidings, similar to those it provided at Rushcliffe, for Shepherd's plaster works.

In the Nottingham area New Basford provided the goods facilities for the City in general although there were two small goods handling areas at the north and south end of Nottingham Victoria. Also in and around the Nottingham area there was a multiplicity of sidings connected with collieries and coal traffic such as at Hucknall, Bagthorpe and Bulwell Common.

K3 class 2–6–0 No.61839 on a Banbury fish train at Rugby, 24th July 1952. *(J. B. McCann)*

Left: No.1046, 0–6–0 class 9J on an 'up' goods working near Leicester carrying coal for Abbey Lane coal sidings.
(Real Photographs)

Left, bottom: A glorious view of 9F 2–10–0 No.92095 passing through Nottingham Victoria on a Woodford-Annesley freight at 10am on Saturday 1st June 1963.
(T. Boustead)

As far as destinations were concerned the Great Central thrived mainly on its ability to allow traffic to have a continuous run to its destination without being interrupted on its way. It was in fact a major artery which connected several different routes together. Trains could travel freely from the North East to the South West or Wales without any difficulty at all. Wagons could be transferred with ease and sorted at various yards along the route to ensure they went to the right places. During the morning, trains came down from York to Woodford and from there they would go on to Bristol or Cardiff, carrying a variety of merchandise. Afternoon and evening trains were mainly comprised of fish workings coming from Hull, New Clee and Grimsby and these were rushed through to Plymouth, the South Midlands and South Wales.

Annesley, Woodford, Banbury, Leicester and York supplied motive power for the freight workings. In the early years locomotives used on these runs included 'tinies' or 04s, B5s and Atlantics which later gave way to V2s, K2s, B16s and B1s. As time went on even the Britannia class appeared on fish trains. Another noteable working was the parcels trains which were run between Nottingham Victoria and Marylebone, and Banbury and York. An additional tradition in later years was to add parcels vans to the afternoon slow Nottingham to Marylebone passenger train.

The Annesley to Woodford freights were run to a tight schedule and over 40 trains were able to

With the sun beating down through the canopy and steam everywhere, K3 class 2–6–0 No.3815 coasts through a deserted platform 10 at Nottingham Victoria, 8th May 1946 with train of mineral wagons for Woodford Halse. *(H. C. Casserley)*

A1 class 4–6–2 No.60125 Scottish Union *on Saturday 27th June 1964 with the return leg of a Normanton to Didcot pigeon special nearing Bagthorpe Junction, which lay between New Basford and Bulwell island platforms. A1s were rare performers on the London Extension.* *(T. Boustead)*

be run in either direction. Each engine ran up to Woodford with a loaded train before returning home with empty wagons for the next full run. By 1950, when the service had only been in operation for a period of three years, some 680,000 wagons had already been exchanged with the Great Western at Banbury. The link line could no longer cope so the Stratford upon Avon and Midland Junction Railway was also exploited.

At first the Annesley to Woodford trains were worked by ex Great Central or London & North Eastern engines, but in the late fifties the British Railways 9F class engines were drafted in and their performance revolutionised the intense freight service. In speed the trains were limited to 30mph for loaded and 33mph for empty trains, though drivers often put the engines through their paces. Speeds of 50 or 60mph were not unknown with the result that it was usually possible to fit in three round trips in one day. The drivers nicknamed the trains 'Windcutters' and 'Runners' and not without good reason as they were an elite service that was a paragon of efficiency. Their removal in 1965, along with all the other 66 freight services in 1965 is in some ways beyond comprehension.

Right, top: K3 No.61925 photographed on 29th May 1959 whilst working a train of fish empties from Banbury between Bulwell and Hucknall Central. *(T. Boustead)*

Right: Class 9F No.92088 on a rather dull day in April 1959 working an incredibly long 'down' freight for Annesley through Kirkby Bentinck. *(T. Boustead)*

61925

Private Sidings connected to the Great Central London Extension

Calvert Brickworks	These sidings were provided to the south of Calvert Station the first station on the London Extension going north, for Messrs Itters Brick Company which later became part of the London Brick Company. The sidings survived in use until 1977. The nearby LNWR. Oxford-Cambridge line also had sidings for the brick traffic.
Charwelton Ironstone Company	Traffic from these sidings was always very heavy right up until closure of the goods yard. Wagons loaded with the mineral came from a railway whose access was made via a gate at the north end of the goods yard.
Staverton Siding	Wayside Road for wagon storage.
Barby Sidings	The sidings here, which lay on the south side of Rugby Central, were erected to serve a war time supply depot, and to control traffic in and out a signal cabin was built.
Rugby North Sidings	Near the site of the LNWR at Rugby the Great Central provided a siding, two small shunting necks and a loop which served a cattle dock, which also had its own signal box. These though had disappeared by the 1960s.
Newton Siding	Served a gravel pit to the north of Rugby.
Shawell Siding	Wayside Road for wagon storage.
Abbey Lane Sidings (Leicester)	Used for coal traffic and distribution of the latter by the Great Central and later British Railways.
Swithland Sidings	Exchange and sorting area for the Mountsorrel granite quarry railway.
Loughborough South	Short spur to a nearby wood, used for timber traffic.
Barnston Sidings	Constructed for the Barnston Blue Lime Co. for the transportation of gypsum. Whilst the 'up' line sidings have been removed those on the 'down' side still remain in use today.
Gotham Branch and Sidings	Served a series of gypsum mines owned by a Mr J. W. Shepperd. The line ran for $2\frac{1}{4}$ miles and contained a passing loop, while the sidings like those at Swithland were put close to the main line to enable exchange of wagons to take place.
Ruddington Railway	This was a later addition built for the war effort and later used as an ordnance depot. The military railway made a connection with the Great Central at the south end of the goods yard. Traffic continued to flow from here until 1978.
Wilford	Situated on the outskirts of Nottingham a couple of sidings were provided here for a small brickworks.
Clifton Colliery	Branch leading off the main goods yard to the south of Nottingham to a colliery.
Bagthorpe & Bulwell	Receiving and holding sidings were provided at both these points as a major junction lay here. The same was true of Woodford Halse which afforded many connections to the south.
Hucknall	A short branch to a colliery was provided here which led off from the south end of the station yard.

Locomotive Sheds

Annesley and Nottingham

At the opening of the Great Central two depots served the area around Nottingham. These were sited at Annesley to the north, and Arkwright Street to the south, supplying freight and passenger engines respectively.

The shed at Arkwright Street was only to have a very short life, handing over its role to Annesley.

Annesley locomotive shed on 12th May 1962 with Southern Region 'Schools' class No.30925 **Cheltenham** *receiving attention after being coaled up with an inordinate amount of coal! Behind is Midland Class 2P 4–4–0 No.40646 which had officially been condemned two days earlier. (K. C. H. Fairey)*

It was equipped with an inspection pit, turntable, coaling stage and water columns, behind which was a shed which was designed to accommodate 16 engines on four roads. Along the side of the buildings were mess rooms, stores and offices, though they were to see little use. Soon after the line opened the shed was closed but the site remained in use as a carriage storage area and goods yard. All locomotive activity was transferred to Annesley which was gradually being expanded as its importance grew and the London Extension attracted more traffic.

Originally the Manchester, Sheffield & Lincolnshire had planned to site its main shed at Bulwell, where there was a large junction complex and the ability to site the depot closer to Nottingham itself. The drawback was that the local corporation demanded much higher rates for this area so it was decided to make Annesley the main shed. However this site had two disadvantages, namely it was deep in the Nottinghamshire countryside, far from any station and also there was a problem regarding mains water supplies. The advantages of Annesley were its proximity to the Great Northern Railway, which had a link with the Great Central and also the fact that the site was within a convenient distance of the large marshalling yards provided for traffic which was received and despatched from the local collieries. There was great scope too for expansion and the Great Central actually owned more land than it used, although it was always intended, that should a settlement appear, then this area would be needed.

Unlike Woodford Halse where the Railway had built its own houses and lodgings for the men Annesley had no accommodation and could draw no workers from the immediate local area. Therefore means had to be provided to bring the employees to the site. The nearest workers came from Hucknall, Bulwell and Kirkby which were all some distance away. As there was no accommodation save for a few huts for railmen on overnight workings, the Great Central provided a train which became a regular feature from the early days, running between the places mentioned earlier. Usually the train was made up of one or two carriages with an engine pulling it that ranged from an express passenger or freight engine to a small tank engine. Examples of engines which did pull the workmen's train were the celebrated 9Fs and B1s in British Railways' days, while in the Great Central days Sacre 2–4–0 tanks and 4–4–0s were used. After the LNER had taken over control of the London Extension a Sentinel steam railcar was brought in to work the turn on occasions. Eventually the railcars proved to be most cost effective and were introduced on a regular basis until they became defunct in 1940. They were ser-

viced at Annesley Loco and cleaned at Basford where they ended up once or twice a week.

The train was operated to a regular pattern, with few exceptions, although after 1925 it only ran the full length of the Great Central route on Sundays, partly using the Great Northern section during weekdays. The number of trains which were scheduled over the route were four and fourteen respectively each way. The train was always busy and apart from one or two alterations or refinements the service remained the same up until closure. In fact it was given the name 'Dido' and there have been those who have tried to think up academic reasons for the nickname in the form that Dido, who was the Queen of Carthage was in some way comparable with Queen Victoria. An alternative was that the start of the service occurred on the day of Queen Victoria's Diamond Jubilee and another that the name may have been a corruption of Diadem which was the name of a local hill. These all seemed very unlikely and after inquiry from a Great Central man in the Nottingham area the real reason for the name was imparted to me. This was because of the regularity of the service ie it ran 'Day In, Day Out', the name 'Dido' was bestowed on the train! At the Annesley end the 'Dido' was given an official pair of setting down points in the mid 1930s with the erection of Hollin Well and Annesley plus a halt at Annesley South Junction which was a very simple structure due to the fact that its use was only intended for railwaymen.

Turning to the actual Motive Power Depot there was a turntable, a coaling stage, water columns, and engine crane and a shed with sidings to accommodate some 70 engines. Attached to the shed building were the usual mess rooms, an engine room, offices, smiths' shops and stores. The total number of men employed on the depot included footplatemen, tradesmen, cleaners, fitters, shed labourers, supervisors and administrative staff, some 500 in all. To most of these men Annesley was a hostile place as there were no facilities socially or otherwise in the locality and the fact that water had to be carted in containers, which were old engine tenders, only added to the discomforture.

Annesley depot always had a new selection of engines from the Sacre tanks and 4–4–0s, D9s and D12s, the Clayton and Sentinel railcars, C14s and Q4s to the O1s and O4, C12, J5, J11 and K3 classes in Great Central and London & North Eastern days. One other type which provided the shed with three engines during the latter period was the B1 class which was to last on shed into British Railways' days when the depot played host to a whole range of different engines. These included Stanier and Fowler ex London Midland & Scottish Railway engines, as well as some which had lasted from the LNER stud to the BR 'Britannias' and 9Fs, famed for their performance on the Annesley to Woodford freights.

A rather misty overall view of the shed with the two visiting engines on the right attracting a small crowd of enthusiasts around them. These engines had worked up to Annesley to be made ready to haul the RCTS 'East Midlander' enthusiasts' special from Nottingham Victoria to Darlington and back.

(K. C. H. Fairey)

Annesley shed with a varied line up of engines. On the left are some of the 53 01 class engines allocated while the locomotive facing the camera, centre is of the K3 variety, leaving those on the right in front of the sheer legs. These are a mixture of J50s, J11s, K3s and more 01s. *(G. C. Museum, Loughborough)*

Leicester Central

The shed at Leicester Central, although regarded as a small shed like that at Nottingham Arkwright Street which closed, was considered very important because it was situated mid way between London (Marylebone) and Manchester. Leicester was unique in that it did not follow the usual criterion for building a locomotive depot, which was that it should be near an important junction, like at Woodford Halse, Annesley, Nottingham and Neasden. While some depots were sited at terminals such as Marylebone and Aylesbury, which was partly considered to be a terminus by the Metropolitan Railway. 'Lester', as it was known by the railwaymen, was mainly concerned with the provision of relief engines for fast train workings and local trains as well as motive power for freight and the odd cross-country working.

The depot was noted for its efficiency and smooth operation which was effected by the vast number of staff that the shed had, some 170 men comprising of 70 pairs of footplate staff while the rest made up the maintenance staff. There were nine fitters, a boilersmith, a tinsmith, two blacksmiths, a turner, a turner-fitter, two boiler washers and twelve mates, as well as the usual coalers, labourers and young men who were starting to learn about engines. The latter staff were responsible for maintaining their locomotives in between trips to Gorton for major work. Leicester not only gained a good reputation for its repair and maintenance facilities but also it had the distinction of drawing men away from other depots on the Manchester, Sheffield & Lincolnshire system. This trend was an advantage as it meant that locals who enrolled at the shed were trained by experienced men.

The engines the crews worked were light weight, powerful performers designed by Pollitt, Parker and Robinson who were the Great Central's main locomotive engineers. When the London Extension was first opened, Leicester Central shed was supplied with three 11A class 4–4–0s Nos 806, 861 and 864 which were to last in service for a number of years to come. A variation on the class was later brought into service known as the 11Bs which were designed by Robinson. They proved very popular and made light work of expresses and later, as traffic grew heavier, so their ability was tested. As well as the 11B class some 9H class 0–6–0Ts for shunting duties were allocated to Leicester. For good trains and short trips the Great Central provided some of the 9F class 0–6–2 tank engines while two of the Sacre Class 15 types were employed on local passenger trains. An American style of this class was also based at Leicester after being at Gorton for a while and these were firm favourites with the crews.

In 1903 the Class 13 singles, which had been based at Neasden for local workings, appeared at Leicester but were only to last here for a year before

42784

Left: *Leicester GC locomotive shed on the 26th February 1959. On the left by the shed wall is a departmental stores van and mess carriage. The locomotive is B1 class No.61298 on the right of which is the sheer legs used for lifting vehicles for repairs to wheels and axles.*

(K. C. H. Fairey)

Left, bottom: *Hughes-Fowler Class 6P5F 2–6–0 No.42784 at Leicester GC Loco. on the 26th February 1959 having just backed onto the turntable. To the left in the background is Leicester power station.* *(K. C. H. Fairey)*

being transferred away. During 1904 Leicester played host to some railcars which were used to operate short haul services to Loughborough in one direction and to Rugby in the other. Although fairly economic, like the ones at Annesley they were liable to failures and in 1919 were transferred away to Lincolnshire where they ended their days.

Atlantics continued to be the main engines at Leicester although new classes appeared frequently prior to grouping, such as the Class 1A 4–6–0s – the 'Glenalmonds' which were built over a two year period. Essentially these eleven locomotives were for mixed traffic but they were found to perform much better when working goods trains than passenger diagrams. Another class of locomotive which was shedded at Leicester was the 9Q which was also a mixed traffic engine. These 4–6–0s were often pushed far beyond their limits to attain high speeds on the fastest trains which were run on the Great Central London Extension. Often the 9Qs were used for excursion work and were seen on various parts of the Great Central system well away from their native depot. Some of these engines were in fact not completed until after the amalgamation of the Great Central into the London & North Eastern Railway who classed them B7. There were three types of engine which were used for shunting and local train operations which are worth mentioning. These were the Class 18 0–6–0 tanks, the Sacre 4–4–0s and finally the 9J class 0–6–0s which were easy to maintain being good, simple and powerful engines and had the distinctive nickname 'Pom Poms'. Aside from goods they saw the occasional duty on local trains and excursions.

Little change occurred at the depot after the grouping of the Great Central until 1930 when the London & North Eastern whose designer was

Leicester locomotive shed on 12th March 1961 with a BR Standard Class 5 4–6–0 No.73157 and two B1 class 4–6–0s present (Nos.61137 and 61380). The whole structure is dwarfed though by the massive power station in the background.

(K. C. H. Fairey)

Two B1s Nos.61269 and 61063 stand simmering at Leicester GC locomotive depot on 26th February 1959. (K. C. H. Fairey)

Gresley (the successor to Robinson who had been the instigator of most of the fine engineering creations which had appeared at Leicester), decided to introduce a new breed of engine to replace the Atlantics which had been in operation for some 30 years. The new type were known as the B17s and had a 4–6–0 wheel arrangement, though again like the former Atlantics, they were lightweight medium power engines, and it was not unknown for Leicester drivers to achive speeds in the 80s and 90s. The engines themselves were known as the 'Footballer' class due to the fact that they carried the names of football clubs.

After the 'Footballers' the last significant development in the locomotive field was the building of the B1 class 4–6–0s by Edward Thompson who was Gresley's successor. These were completed before the Second World War and eleven of this new class were allocated to Leicester but their number at the depot was to grow to thirteen before they were transferred to Woodford and Colwick in 1962, when the closure of the shed was originally intended. When the change in BR region took place in 1958, from the Eastern to the London Midland, ex London Midland & Scottish engines began to appear in the form of Stanier Class 5 4–6–0s including two of the class that were completed in BR days, also arriving were Class 4 Staniers and even some of the Class 3s of Stanier and Johnson. Of course, the London & North Eastern engines did not disappear and even in 1959 there were five of the GC's all hallowed V2 class 2–6–2s on shed. Leicester, also for a short time, played host to a group of engines which boasted a similar performance to the V2 and these were the A3 class 4–6–2s which had eight of its members on shed between 1950 and 1955. At the time of the Modernisation Plan in 1955 no proposals were made for the 'dieselising', though in 1963 one 350bhp shunter was allocated to Leicester until closure took place.

In layout and capacity Leicester was a small depot which was able to accommodate 15 to 20 engines on the four roads which lay under cover. Inside the shed there was a lifting gantry as well as the sheer legs on the outside. On the side of the shed were the usual offices, smiths shop, mess rooms and workshops while outside the shed an inspection pit and coaling stage were provided. The coaling stage unlike others along the London Extension remained unmechanised right up to the time of closure. Other facilities included water columns and turntable plus numerous sidings.

Woodford Halse

The shed at Woodford Halse was provided to supply engines for both goods workings and passenger trains as well as shunting duties around the massive

9F class 2–10–0 No.92215 simmers gently in front of Woodford Halse locomotive depot on the 7th May 1961, with a WD class 2–8–0 behind. *(K. C. H. Fairey)*

exchange yards and repair facilities. One justification for the siting of the shed at Woodford Halse was that the main line formed a junction with the East and West Junction Railway which joined up with many other main lines to the north and south. It was also necessary to provide facilities at a regular distance apart to enable the smooth running of the railway. There is no doubt that a further factor which influenced the construction of a major depot here was the availability of a vast area on which to build it, like at Annesley.

Indeed Woodford was expanded at regular intervals throughout its short career as a shed, the first occasion being when the link to Banbury was opened which brought the presence of Great Western engines whilst they were working on exchange diagrams in the goods yard. In layout the shed had six roads which were able to hold 30 engines and in general the depot had the same facilities as Leicester in the form of a water tank, coaling stage, electric power house, ash pits and of course the usual offices attached to it, including mess rooms, fitters' shops, a shedmaster's office and stores. The whole structure was built out of blue engineering bricks and in the centre, on a small gable, a clock was situated. This was later moved to a position above the shedmaster's office because it used to be regularly blackened by engines entering the shed! The clock was moved after Nationalisation in conjunction with the complete re-roofing of the shed. To the left of the main building stood a set of sheer legs for lifting locomotives, which required minor attention, such as to the bogie trucks or the driving wheels. Situated to the right of the shed, instead of the usual turntable facility, a rather more lavish arrangement for the turning of engines was provided. This was a triangle of track which replaced the turntable in 1941. Closer to the shed lay four sidings that served the locomotive coal stacking ground and a mobile crane was used to load the coal into wagons when it was needed. Locomotives had been laboriously coaled by hand from the beginning of the London Extension's presence at Woodford and the situation was not changed until the 1930s when the LNER had an enormous mechanical coaling plant erected. Like Annesley water was a problem in the early years too and this had to be brought in to the locomotives but such a drawback did not impede the efficiency of the depot. In fact the Great Central was always well prepared for any emergency and in 1947 an impending coal strike prompted the LNER to erect a series of oil storage tanks. These were demolished some ten years later never having been used.

One facility that the railway provided here which was not available at Annesley was housing for the workers, and this was to change the face of the sleepy community at Woodford. Altogether there were some 136 terraced houses which were built near the embankment by the river Cherwell. Four of these houses were converted for lodging purposes for drivers on overnight turns who needed a place to stay. The hostel was opened in 1907 and continued until 1941 when lodging ended. By 1956 the locomotive department used these houses and not the traffic department. The size of house allocated and its situation showed how important its occupant was in the hierarchy of the railway. Those who lived near the embankment were usually lower grade workers while those who lived in houses with bay windows were usually drivers and more important employees. These employees could be on call at any time and in fact the Great Central employed call boys who were usually starting on the long apprenticeship to become an engine driver. These call boys were responsible for walking round the various streets where locomen lived and knocking them up ready for their shifts.

Woodford engine crews handled a great variety of locomotives, probably more than any other depot along the London Extension because of the fact that there was a large amount of traffic which was being despatched and received from all over the country. In the early days the depot played host to the famed Atlantics and Robinson 11A and 11B class locomotives for passenger trains, while J9 0–6–0s were employed for goods. A more powerful design though which saw service from Woodford shed was the 11E or 'Director' class whose excellence and performance kept them going well into the Gresley and Thompson era. The D9 design in fact was actually upgraded twice with the building of the D10 and D11 designs. In the years of the two famed London & North Eastern engineers the stud at Woodford saw the appearance of four of the 'Footballer' class locomotives and the B1 4–6–0s which were not to be confused with the two introduced by the Great Central in 1903/4.

Most engines at Woodford in fact were inherited by the London & North Eastern from the Great Central, such designs including the Atlantics and the B7s which were regularly employed on fish trains. In addition to the latter there were the B5s which became regular performers in British Railways' days. By the Second World War, locomotives which appeared at Woodford were the J10, J11, L3 and N5 types of Great Central locomotives, while

On the 3rd October 1964 a J39 class 0–6–0 No.64747 languishes by the water tank at Woodford Halse. In the background is the coaling plant and the wagons seen to the right of the water column supplied coal for the locomotives. *(K. C. H. Fairey)*

those from the LNER included the J39, K3, L1 and V2 engines as well as those mentioned earlier. By this time Woodford was stabling some 54 engines, 20 more than the shed could house.

After the war and Nationalisation the situation changed so often, where locomotives are concerned that it is necessary to use a table to show which classes were allocated over the last ten years. As one can see Woodford was blessed with the presence of four diesel shunters of the Class 08 variety although no main line diesels were ever allocated to Woodford. Originally it was hoped that the depot would play host to a small number of diesel locomotives for express work and British Railways even went to the lengths of building a diesel depot and installing another couple of fuel tanks which again remained unused. It was about the time the diesel facility was being built on the side of the main shed that the depot came under the control of the London Midland Region.

Locomotives allocated to Woodford Halse over the last ten years.

Class	Type	1955	1959	1963	1965
LNER					
B1	4-6-0	3	5	2	–
J11	0-6-0	8	–	–	–
L1	2-6-4T	1	3	–	–
L3	2-6-4T	1	–	–	–
V2	2-6-2	6	3	–	–
WD	2-8-0	25	26	23	–
K3	2-6-0	8	–	–	–
LMSR					
4	2-6-0	–	2	–	–
3F	0-6-0	–	3	–	–
8F	2-8-0	–	–	3	4
5	4-6-0	–	–	5	1
4	2-6-4T	–	–	7	1
BR					
4	4-6-0	–	–	3	–
5	4-6-0	–	–	7	–
08	0-6-0 DE	–	4	4	–

The last six locomotives on the roster at the time of closure in 1965 were Nos 42082, 44835; 48002; 48010; 48035 and 48121.

Neasden

Neasden was the most southerly of the depots on the London Extension and was situated opposite the large depot which had been established by the Metropolitan some years earlier. The shed was some five miles from the London terminus of Marylebone, and like Woodford had capacity for 30 engines on a six road basis. All the usual facilities were provided here as at the other depots, and like that at Woodford there was a small colony for railwaymen who worked at the depot. Neasden also had its own lodging facilities which were mainly used by men working turns from Gorton depot in Manchester.

Neasden was responsible for the supply of engines for the Bradford and Manchester expresses and these were usually in the form of the Atlantics and the 11A and 11B class during the first few years, giving way later to the 'Director' class which gave stalwart performances on the Great Central. One 'Director' based at Neasden rivalled its more modern successor the 'Footballer' class by completing 14 round trips to Manchester as opposed to the 10 allowed for the B17. As well as being blessed with a variety of engines from the D11 and D10 classes, Neasden also stabled a variety of B7s known as 'Black Pigs' which were in effect one of the last types to be built by the Great Central before Grouping. These engines were never really regarded highly as they were merely an upgrading of another class which was shedded at Neasden for a while, namely the B3s. Known as the 'Lord Faringdon' class these worked alongside the 'Directors' on Manchester trains until the arrival of the B17s which took over some of the diagrams.

As at the other depots along the London Extension the pattern was ever changing as far as locomotives were concerned with new types appearing quite frequently. These included A1s, B1s and later many ex LMSR types and British Railways' Standard designs which are shown in the accompanying table. Apart from supplying engines for express workings Neasden was also responsible for the provision of locomotives for local trains to Princes Risborough, High Wycombe and Aylesbury. The Verney Junction and Chesham branches were also worked, being shared with the Metropolitan.

Suburban traffic was always heavy and in this respect Neasden was always a very busy depot. However for locomotive running purposes Neasden's official area ran via both Metropolitan and Great Western routes as far as Woodford Halse which was the next main shed going to the north. Subsheds were provided at Aylesbury and Marylebone though the one at the latter closed before Grouping and was only ever a single road affair, while that at Aylesbury was predominantly used by Great Western engines.

The mainstay of the local trains as far as locomotives were concerned were the A5, F1 and F7 tanks although at the beginning, until Grouping some of the small Sacre tanks were used on

2–4–0 No.362 along-side Neasden shed in 1906 with an auto train. *(V. R. Webster)*

Valour *No.6165 on shed at Neasden, 22nd August 1937.* *(L. Hanson)*

Fairburn Class 4 2–6–4T No.42222 at Neasden M.P.D., 22nd April 1954. (Author's collection)

push-pull workings. Other types which became acquainted with London inner suburban workings were the C13 and C14 breed of locomotives whose classification had been 9K and 9L respectively in Great Central days. In fact the A5 and the other tank engines mentioned above completely took over and most types had vanished by 1925, leaving the A5s to soldier on, well into British Railways' days.

One other feature made Neasden a very important depot and this was the construction of a loop line around the newly completed Wembley Stadium, which had none of its own engine facilities. Therefore trains working on specials to Wembley usually travelled down to Neasden for cleaning, coaling and watering ready for the return journey. Traffic was particularly heavy during the British Empire Exhibition in 1924, such that extra engines from all parts of the LNER system were called in to help with the trains and stock manoeuvring. Such types included a Y7 0–4–0T from The Dock in Newcastle, Great Eastern tram engines from Wisbech and a J62 from Staveley depot. After the end of the exhibition at Wembley, Neasden became responsible for providing locomotives for a newly opened branch to Watford which was agan shared with the Metropolitan. These services were carried out by the new L1 tanks which had originally been among those drafted in to help with the Wembley Exhibition traffic and as one can see from the table, they were still very much in evidence in British Railways days.

Allocations of locomotives to Neasden Depot over the final twelve years to closure in 1962.

Class	Type	1950	1955	1959	1962
LNER					
A3	4–6–2	3	3	–	–
A5	4–6–2T	5	–	–	–
B1	4–6–0	6	10	5	3
C13	4–4–2T	2	3	–	–
J11	0–6–0	3	–	–	–
L1	2–6–4T	37	27	–	–
L3	2–6–4T	6	–	–	–
N5	0–6–2T	11	7	3	–
N7	0–6–2T	5	–	–	–
Y3	0–4–0T	1	–	–	–
GWR					
1400	0–4–2T	1	1	–	–
6100	2–6–2T	2	–	–	–
LMSR					
4	2–6–4T	–	11	28	–
BR					
4	2–6–4T	–	10	10	–
4	2–6–0	–	10	10	–
08	0–6–0DE	–	–	4	4

Aside from the locomotives mentioned above at closure the three B1's which are reputed to have remained until the fateful date in August 1962, were supplemented in the last six months by a fleet of the 'Britannia' class which had been brought in to work trains transferred to Marylebone during the Euston electrification work. These were later sent up to Annesley after Neasden shed closed.

Locomotives

Prior to opening in 1899 the Great Central employed goods engines on the coal trains that ran on the London Extension. The purpose of these trains was to ensure the level had settled and that the structures en route, such as the viaducts at Bulwell, Nottingham, Leicester, Rugby and Brackley would stand up to the use they were going to experience. When the first trial passenger train ran on 7th November 1898 a Pollitt Class 11A 4–4–0 was used to run from Manchester London Road to the then incomplete Marylebone station. The Class 11A engines were an improvement on the Class 11 engine design by Parker and were brought out in 1897 in anticipation of the opening of the London Extension, by Harry Pollitt who had become CME of the Manchester, Sheffield and Lincolnshire Railway in 1894. The latter engines were built by Beyer Peacock of Gorton and saw service on Marylebone to Manchester Expresses. Minor modifications were made to these engines over the years by Robinson who was Pollitt's successor and who added chank pin splashers. All these engines later were relabelled Class D6 as they survived into LNER days.

Another pair of classes which performed stalwartly on passenger services and goods alike were the Pollitt 9G class 2–4–2Ts of 1898 which varied from Parker's Class 3 in that they had a slightly higher set boiler. Instead of the Joy's valve gear they had a link motion, while the other maid of all work was the 9F class 0–6–2T (class N5 of the LNER), of which some 125 were built from 1899 onwards. Members of the former class saw service on the London Extension, primarily in the London and Aylesbury area while the 9Fs were mainly confined to shunting duties or working local trains – these engines being more common on the northern sections of the Great Central. One class of locomotive, the 9H 0–6–0 was built specially by Beyer Peacock for the London Extension goods trains. The 9Hs were originally designed by Harry Pollitt while John G. Robinson continued their construction until the emergence of his famed J11 'Pom-Pom' goods engines. Shortly after the 9Hs came Pollitt's last design, the Class 13 single, six of which were built at Gorton in 1900. One engine, No.969 which appeared soon after Robinson became CME, was

turned out in an experimental darker green livery and was fitted with rectangular number plates. The Class 13 singles saw four years service on the London Extension before they were all transferred away to the Cheshire Lines in 1904.

Shortly before the singles were introduced a series of engines known as the Class 15s were ordered from the Baldwin Locomotive Works in America and at the onset these locomotives were branded 'Yankees', and saw limited service on the southern section.

A major change began to take effect in 1900 when the specifications and general bodywork of the motive power built for the Great Central London Extension was updated as a result of Robinson, who became the Company's new locomotive engineer. His locomotives soon became known for their sleek, stylish looks and improved performances over the previous designs of Harry Pollitt. The first of the new breed of locomotives were the 9J class 0–6–0s of 1907 built by Neilson. Because of the raucous bark of their exhaust whilst working goods diagrams, the 9Js became nicknamed 'Pom-Poms' after a rapid firing gun which made a similar noise. As with the 11A passenger class several improvements were made to later builds with the incorporation of enclosed four column safety valves on the engine and plates on the top of the tender sides instead of coalguard rails.

In the same year as the 9J class was introduced John Robinson designed his first passenger locomotive based on the Class 11 series of Pollitt. The 11Bs, as the engines became known, appeared in 1901 and were smart looking powerful workhorses admirably demonstrated when, at the end of 1903, No.1040 ran the $126\frac{1}{2}$ miles from Marylebone to Nottingham in 123 minutes, hauling seven coaches and forming the 10am express. Two rebuilds took place in 1907 with Nos 104 and 110 being converted to take a larger 5ft diameter boiler and were redesignated Class 11C. Neither of the latter were used on the London Extension, though the 11D, a further improved version of the series saw service on many and various workings on the Great Central London Extension from the Newcastle to Swansea, Sheffield to Bournemouth and York to Bournemouth trains as well as local Nottingham to Rugby trains.

A break from the conventional Class 11 design and the 'Pom-Poms' took place with the introduction of the Class 8 4–6–0 mixed traffic engine which was adorned in the goods engine livery of black, lined out in red and white. Class 8s were as equally capable of hauling passenger trains as they were goods trains. They mainly became noted for pulling fast fish trains and gained the nickname The 'Fish' class. A departure in design was made soon after the series was introduced with the creation of the 8A 0–8–0 engines, mainly used for mineral work. A further improved version, the 8B 4–4–2 was to become a ubiquitous work horse mainly because of

Left: A J11 (GCR 9J) class 0–6–0 No.5223 awaits the all clear with an 'up' goods working at Nottingham Victoria, 7th June 1927. ***(W. L. Good)***

LNER C4 class 4–4–2 No.5365 **Sir William Pollitt** ***sitting in the south goods dock at Nottingham Victoria, 1st June 1929.*** ***(W. L. Good)***

their large number. There were 25 of this class all told, incorporating the saturated and superheated engines. The 8Bs (known as the Atlantics), in addition to being LNER class C4 were employed on most services from the Manchester Expresses to the long distance cross-country workings. This series was further improved with minor additions resulting in sub classes C, D, E and F being evolved. The 8D and 8F types saw much service on the London Extension on all services.

While the Class 8's were fulfilling the need for more powerful engines as the number of long distance services increased, the Great Central realised there was a market for introducing more suburban services which would require their own kind of motive power. These services were catered for with the 9K and 9L tanks of 1903, both of which had the 4–4–2 wheel arrangement. The 9K was used on a number of services from semi fasts between Nottingham and Marylebone to short hauls on the northern section of the London Extension. The twelve members of the heavier 9L version, also built by Beyer Peacock, were put into service on the Great Western and Great Central Joint line.

LNER C13 (GCR 9K) 4–4–2T No.6065 at the south end of Nottingham Victoria on a Rugby local working.
(D. J. Montgomery)

For freight traffic the Great Central turned out a heavy engine with a 2–8–0 wheel arrangement known as the 8K, which proved very useful during

2–8–0 No.5 originally built for government use in World War 1 but later taken into GCR stock as class 8K. They later became LNER class 04, this locomotive being renumbered 5005.
(Real Photographs)

the First World War. Following the construction of the latter a new variety of 4–6–2 tank engine was designed and built as the Class 9N, the first ten of which were designated for the London area. Due to their success a further six were constructed a year

Ex GCR 9N class (LNER A5) 4–6–2T No. 69803 at the south end of Nottingham Victoria in 1948 by the platform exits to Parliament Street. (D. J. Montgommery)

later. During the same year passenger loadings dictated a need for a new type, the Class 1 4–6–0s or the 'Sam Fay' class, as No.5423 was named after Sam Fay, General Manager of the Great Central.

These engines displayed a number of interesting features in that they had the largest inside cylinders

GCR Class 1 4–6–0 No.423 **Sir Sam Fay** ***at Annesley shed. This locomotive and the rest of its number were classified B2 by the London & North Eastern Railway. With the appearance of the 2 cylinder rebuilds of the 'Sandringhams' the engines were further reclassified. (Photomatic Ltd)***

LNER B8 class (GCR 1A) No.5280 at Nottingham Victoria in the centre road on 7th June 1927.
(W. L. Good)

LNER B8 class (GCR 1A) No.5004 Glenalmond *on a Leicester passenger at platform 11 at the south end of Nottingham Victoria.*
(D. J. Montgommery)

LNER D10 class (GCR 11E) No.5432 **Sir Edward Fraser** *at Nottingham Victoria in 1937.* *(D. J. Montgommery)*

of any kind of locomotive, they also set a precedent as far as having continuous splashers and nameplates were concerned. A goods version of the engine was built a year or two later, although these locomotives which became known as the 'Glenalmonds', differed in appearance and specification.

In 1913 a further type of locomotive was introduced by the Great Central which was specifically intended for heavy passenger work. These were known as the 11E 'Director' class and all ten engines were named after the members of the Great Central Company's board of directors.

The class 11E engines were so successful that a further batch known as the 'second series' or 'Improved Directors' was built, consisting of eleven locomotives known as the 11F class which were outshopped between 1919 and 1924. All were named

LNER D11 class (GCR 11F) No.5504 **Jutland** *at Nottingham Victoria of the 1st June 1929 with a Manchester to Marylebone train.* *(W. L. Good)*

L N E R
5510

VALOUR
6165
L N E
LNER
3

and were the last locomotives to be built before the Great Central became part of the LNER at Grouping in 1923.

Left: LNER D11 4–4–0 No.5510 **Princess Mary** *and C4 class 4–4–2 No.6090 at Leicester Central on 7th August 1937.* *(N.G. Steele)*

Another class, which was probably one of the finest and best performers of Great Central passenger locomotives, was the 9P built towards the end of Great Central ownership of the London Extension. These engines incorporated a series of interesting variations in design during their working life, such as the fitting of some with an oil burning system during the coal strike of April to June 1921. One of the engines named *Valour*, No.1165 was a moving memorial to the men who died for their country in the 1914–1918 war.

Left, bottom: LNER B3 'Lord Faringdon' class (GCR 9P) 4 cylinder 4–6–0 **Valour** *on Nottingham Victoria's middle road at the north end. No.6165 was named in honour of the company's employees who died in the 1914–1918 war.*

With Grouping in 1923 and old age a prevalent factor in the resignation of John G. Robinson as locomotive engineer, the young Nigel Gresley took on the task of designing and building locomotives until 1941. Several engines had been inherited from the Great Central which were subsequently renumbered, reclassified and decked out in LNER livery. In addition to the former Great Central engines several new types found their way on to the GC London Extension expresses such as the B17 class, the V2 class and the A3 Pacifics introduced in 1928, 1936 and 1927 respectively. Two other designs also appeared, namely the K3s and J39s which were essentially mixed traffic locomotives. The B17s were well known for their performances on Manchester to Marylebone turns. The latter replaced the C4 class Atlantics which had become life expired and their operation was centred on Leicester shed. As well as the Manchester diagrams the B17s worked a number of Bradford turns together with a few workings on the north to west trains.

The A3s had worked at first on the Great Northern line out of King's Cross until superseded by A4s. During 1938 A3s drifted on to the London

LNER B17 class 4–6–0 No.2864 **Liverpool** *at Leicester Central on 13th April 1937.* *(N.G. Steele)*

LNER Class A1 4–6–2 No.4470 **Great Northern,** *testing the new 70ft turntable at Leicester Central on 16th August 1938.* *(N. G. Steele)*

Extension and worked Manchester and Bradford trains as well as the odd York to Woodford freight train. In 1939 the A3s were joined by the V2s which handled most services with ease. That year also saw the outbreak of war which brought about a complete withdrawal of passenger expresses. In the early part of the war more Pacifics were employed on the London Extension and were the mainstay of motive power in a period when the services had been drastically re-organised. As a result of this several types of locomotives which had been in service up to 1939

A3 class 4–6–2 No.60039 **Sandwich** *at Leicester GC locomotive depot on 24th February 1957.* *(K. C. H. Fairey)*

An exhibition of LNER engines on 13th May 1939 in Leicester goods yard with A4 4–6–2 No.4468 Mallard, *B17 4–6–0 No.2865* Leicester City *and V2 2–6–2 No.4825.* *(N. G. Steele)*

were withdrawn having become life expired.

Little attention was paid to the construction of new locomotives for the London Extension save for the introduction in 1942 of the Class B1 4–6–0 designed by Thompson who had succeeded Sir Nigel Gresley as Locomotive Engineer. These engines were to provide stalwart service for a period of over 20 years. When Nationalization took place in 1948, the majority of express trains were in the care of the B1s shedded at Neasden, Leicester, Annesley and Woodford Halse. Two years after British Railways was formed Class A3s appeared on the London Extension again following a short break during wartime. This time however the Pacifics were limited to working between Marylebone and Sheffield while the B1s tackled the Sheffield to Manchester section of the Marylebone to Manchester diagram. Suburban services in the London area were handled by the L1 class which was introduced in 1945 by Edward Thompson while other turns were worked by the A5 'Coronation' tanks and the C13 class 4–4–2Ts. Two of these were based at Neasden for working the Chesham branch. As time went on Neasden received twelve Fairburn, LMSR 2–6–2Ts for outer suburban work which usually meant trains to Aylesbury, Brackley or Woodford Halse. Woodford Halse was the centre of freight operations on the London Extension and these were mainly handled by 01s, 04s, K3s, J39s and J11s.

Class 4MT Fairburn 2–6–4T built in 1945, No.42250 stands on shed at Woodford Halse, 29th November 1964. *(K. C. H. Fairey)*

In 1958 the London Midland Region took over control of the London Extension from the Eastern

63817

44858

Left: Ex GNR Class 01 2–8–0 No.63817 at Leicester GC locomotive depot in front of the sheer legs, 22nd August 1962.
(K. C. H. Fairey)

Left, bottom: Two days prior to the closure of the London Extension as a trunk route, Class 5 4–6–0 No.44858 stands in the 'up' platform at Brackley on 1st September 1966 at the head of the 5.15pm ex Nottingham Victoria to Marylebone semi-fast. The locomotive seems to have attracted some attention as it is replenished with water. *(K. C. H. Fairey)*

Region and was the prime reason for the Great Central's demise as a main line. The previous year saw the departure of the A3 Pacifics, leaving the V2 class and the B1s on express duty though as time went by the latter were replaced by Stanier Class 5 4–6–0s.

The 'Black Fives' along with the V2s and B1s were employed on the London Extension on 'The South Yorkshireman' and 'The Master Cutler' workings which were the line's crack expresses, though they ceased to operate in 1958 after a period of eleven years. In the same year the Great Central saw the introduction of the large BR 9F class 2–10–0s which were used to haul freight trains, in

BR Standard 9F class 2–10–0 No.92087 on a north bound freight near Belgrave & Birstall, August 1958.
(G. C. Museum, Loughborough)

particular the Woodford 'Windcutters' and Annesley 'Runners'. As well as being used for freight these splendid engines were used to cope with extra seasonal traffic as well as filling in turns on Manchester trains. Two years after the introduction of the 9Fs and the withdrawal of the named trains the last passenger trains departed from Marylebone to Manchester in the hands of a motley selection of class fives.

Up until 5th January 1960 expresses had been the charges of the B1s while the V2s had bowed out the previous year after a fine record. With the removal of express trains, a variety of engines ranging from Class 5 types to LMSR 'Patriots', 'Jubilees' and 'Royal Scots' took on the task of

Below: Class 7P Rebuilt 'Jubilee' No.45735 Comet *is seen passing Swithland sidings on its way back to Annesley depot, running light engine.* *(M. D. Marston)*

Bottom: On 8th August 1964 Rebuilt 'Royal Scot' No.46122 **Royal Ulster Rifleman** *is seen passing Belgrave & Birstall with a relief express.* *(M. D. Marston)*

hauling the Nottingham–Marylebone semi-fasts in addition to the few remaining B1s. Most of these engines saw service on fish trains and newspaper, parcel and pick up freight services, while the 9Fs catered for the coal, steel and heavy trains. Another type of freight engine still very much in evidence were the ex War Department 2–8–0s of which 23 were still stationed at Woodford in the summer of 1962.

Apart from the standard types of engines, there was one experimental locomotive produced by English Electric, akin to a cross between a diesel and a steam train, that is to say the body which was shaped like that of a diesel was mounted on a steam engine chassis and coupled to an adapted standard tender. After exhaustive tests at the Rugby testing station in December 1960, the GT3 as it was known began running on trials in the first week of 1961, between Leicester and London Marylebone. In design the engine was a totally new concept with a gas turbine which was located under the main hood while the exhaust was carried out through two holes in the roof. There were also two large grilles containing filters and above these lay the exhaust for the ejector and oil cooler. The splashers covered up the large battery racks and above the buffer beam there was a fan housing. Inside, its cab was carpeted and there was a generous amount of space for the crew. In livery it was finished in a chocolate colour with orange bands. After a fairly short career GT3 was eventually scrapped.

The closure of Neasden shed occurred in 1962 and with it the transfer of four of the 'Britannia' Pacifics to Annesley which had been transferred to the Great Central to cope with the Nottingham semi-fast trains as steam was gradually being phased out elsewhere.

The stay of the 'Britannias' was short as they were gradually replaced by 'Royal Scots' in 1963 and later Class 5s, until the railway ceased to operate as a trunk route in 1966. A reduction of local services took place in 1963 leaving only the Rugby–Nottingham, Derby–Nottingham and Nottingham–Grantham diagrams. One series of services that remained unaffected were the cross-country workings such as the York–Bournemouth and overnight mail trains between York, Sheffield, Swindon and

A rather unkempt 'Britannia', No.70014 **Iron Duke** *miraculously still carrying its nameplates, draws into Nottingham Victoria on 28th July 1962 with the 4.38pm ex Marylebone semi-fast. Up until the previous month this engine had been based at Neasden until closure when it was transferred to Annesley. The 'Britannias' saw only a year's service on the Great Central before being displaced by the 'Royal Scots'. (Author's collection)*

Bristol running on Sundays until 1963, when Sunday operating over the line ceased. In the summer months the cross-country trains proved even more popular with a number of services working from Leicester to Ramsgate, Derby to Ramsgate, Nottingham to Portsmouth, Manchester to Hastings, Bradford to Poole, Sheffield to Cardiff, Sheffield to Bournemouth, Newcastle to Swansea, Leicester to Scarborough, Leicester to Cleethorpes and Nottingham to Blackpool. Nearly all these trains changed engines at Leicester Central which sported a good variety of motive power on these occasions. Most of the trains from the West Country employed Great Western 'Hall', 'Grange' and 'County' classes south of Leicester where they handed over to the B1s, Class 5s 'Royal Scots' and other ex LMSR types for the northern legs of their journeys.

A spot of Western on the Great Central London Extension was not an unfamiliar sight and here 'Hall' class 4-6-0 No.6911 **Holker Hall** *is seen passing Belgrave & Birstall with a Bournemouth-Bradford express.* *(M. D. Marston)*

A rather clean looking Doncaster designed Class 4MT locomotive No.76052 stands by the water column in front Leicester's engine shed, 22nd August 1962. (K. C. H. Fairey)

On the freight side services continued until 12th/13th June 1965. Over the years since Nationalisation the Great Central saw many goods and mixed traffic engines, from the V2s, K3s, B1s and B16s to the Class 4s, 5s, diesel shunters and even the 'Britannias' which were used on fish trains. Following the withdrawal of freight trains, the London Extension semi-fast and cross-country services survived for another year until 3rd September 1966 – the majority of these being hauled by Class 5s. At closure of the Aylesbury–Rugby section, the line had seen little evidence of diesel traction save for a Class 115 multiple unit which worked the 8.38am Nottingham from Marylebone, the odd Type 3 and the occasional Type 4 on cross-country workings. Between 1966 and 1969 the mainstay of traction over the line's remnants was a couple of diesel multiple units which plied to and from Nottingham and Rugby Central until the service had gone completely on 3rd May 1969.

Chapter 4

Gone Completely

The Great Central, unlike many other main railway routes between the major industrial centres of England, was to be the victim of a period where the mode of transportation was changing over to the motor vehicle, and of the plan which was brought about to reshape British Railways. Many people say that the line arrived too late to be of any use for the places it served which were of any significance, because other railway companies had already catered for the traffic created. It must not be forgotten though that when the Great Central was built each route was owned by a separate company and this competition would be the sauce for keeping the companies in business. However at the time of Nationalisation the state controlled British Railways would find themselves stuck with two or even more lines which served the same places.

An obvious example of this was the London to Nottingham route. Granted, there was unnecessary duplication between the Great Central and the Midland in the eyes of British Railways, but it seems strange that the choice of closure should have been with the Great Central route which was a far better line than that of the Midland as far as permanent way and rolling stock was concerned. The original justification for another line to London had been to provide a modern railway which would cater for new traffic and help ease the burden on the older lines which had already been in existence for 40 years or so. As a line the London Extension was built with economy in mind with small, island platforms for the country stations and slightly grander affairs for the cities, but it boasted a fine selection of purpose built depots and goods facilities and there was also provision made for expansion. The biggest disadvantage the route had, and, this is one of the main reasons for its choice for closure, was the fact that it passed through too many tiny villages and not enough major centres. However it proved a very useful line as far as cross-country traffic was concerned and there was seldom any respite from the clatter of trains.

One may ask at this juncture why should such a line have closed? Well there is a simple answer; it closed because it was starved of traffic and shown to be an economic millstone with the result that the line was no longer justifiable. It is true that the M1 Motorway was partially responsible for its demise but it would be wrong to assume that given the right conditions the Great Central could not have competed with the motor car as railways do today. Traffic which appeared on the roads after the freight service withdrawal, was the result of the need for industry to find a more efficient means of transportation than had been provided by the railways, like the Great Central. Investment at the time was being poured into motorways to the detriment of the railways.

In a sense the years 1955 to 1960 were a period of major transport investment, firstly with the Modernisation Plan for British Railways in 1955, under which, a massive programme of electrification and dieselisation was planned. It was originally intended that the Great Central was to be designated a secondary main line and was to be fully dieselised, though in reality save for the fuel depot at Woodford Halse, a modern station at Banbury, the allocation of four diesel shunters to the line and the occasional passage of diesels on specials working cross-country or filling in turns, the line remained steam to the end. By 1958–59 with the completion of the M1 Motorway and its opening it was clear that the emphasis had been shifted away from investing in railways, and in particular the Great Central route between Aylesbury and Sheffield. This same year also brought about two other significant changes

A quiet day at Woodford Halse in the summer of 1959 shortly after the renovation of the footbridge and rebuilding of the platform on the left, which had been a wooden structure until a concrete surface replaced it. (Author's collection)

for the Great Central, namely the transfer of the two named trains 'The Master Cutler' and 'The South Yorkshireman', both of which had only been running for ten years, to King's Cross. The other change was the transfer of the line to the control of the London Midland Region from the Eastern Region, and thence it became little more than a diversionary route while all attention was focussed on the electrification of the Region's main concern – the West Coast Main Line.

The first major signs that closure was eventually going to occur did not actually appear until 1960, when the Marylebone to Manchester expresses were axed and replaced by a rather indifferent service of three semi-fast trains per day in each direction between Marylebone and Nottingham. With the initiation of the latter the number of passengers began to dwindle as the Great Central happily trundled on taking three hours between the two cities while other routes were being spruced up. Not only did the frequency suffer, but also the quality of the services as refreshment facilities were removed and also the accommodation now was more limited. By 1960 also, the High Wycombe route had lost all its through trains to the north since all the semi-fasts were routed via Aylesbury. Meanwhile the Great Central had lost four of its stations prior to the Beeching cuts, namely Eydon and Chacombe Road platforms which closed in 1956, Braunston & Willoughby, and Culworth which closed to passengers in 1957 and 1958 respectively. There can be no doubt that the two platforms were closed because they attracted little revenue due to the fact that they lay some four or five miles from the villages they were supposed to serve. Although closed to passengers, Braunston & Willoughby, and Culworth remained as goods collection points for a little while longer.

After the shake up of the daytime expresses in 1960 the following year was to provide little change in the operation except for the fact that the line displayed a variety of different classes of locomotive from all parts of the system. This mainly happened as a result of the change in regional control in 1958 and such engines as 'Royal Scot' and 'Jubilee' classes became commonplace alongside the V2s and B1s.

As far as the supply of engines was concerned Annesley was responsible for the north and by 1961 it had taken on much of the duties which were once Leicester's perogative. Leicester shed was run down very rapidly and by the turn of the decade was only being used as a relief depot which provided crews for local trains and the odd cross-country workings. Leicester's main role of course had ceased with the

On a foul wet day in 1964 Rebuilt Bulleid 'West Country' Pacific No.34042 **Dorchester** *heads northwards out of Leicester with a relief Poole-Bradford working, comprising of Southern region stock.* *(M. D. Marston)*

withdrawal of the Marylebone to Manchester and South Yorkshire workings.

At the southern end Neasden depot saw the influx of a number of the 'Britannia' class 4–6–2s which were specifically to work mails, semi-fasts and the sleeper trains which had been transferred over from Euston during 1962 while work began on electrification. The 'Britannias' were kept in a poor state and this reflected the Great Central drivers' attitude towards them. The men mourned the loss of such engines as the V2s and the B1s which, up until 1958 had formed the backbone of Great Central express workings. Quite apart from this the 'Britannias' were already in a run down state. These were only to last for a short while at Neasden shed as the writing had been on the wall for a long time that the depot would shut. By August 1962 all the 'Britannias' had been transferred away to Annesley where they were regarded with esteem and treated accordingly by their crews who cleaned them up and endeavoured to get the best performance out of them.

All this work was to no avail, since the rapid plan of dieselisation and the decision to run down the Great Central meant that the 'Britannias' were to be transferred to live out the rest of their working days in some other outpost. Annesley now played host to the 'Royal Scot' class which for years had been the mainstay of the West Coast Main Line expresses and by the time they reached the Great Central they were old, noisy, unreliable and rather unpleasant to work on. They were employed on the semi-fasts, mails and other trains right up to the end in addition to the odd diesel or V2 which found their way down London Extension metals. The diesel

In the autumn of 1962 an Annesley based 'Britannia' coasts out of Rugby Central past the signal cabin and the entrance to the goods yard sidings and locomotive shed. *(Author's collection)*

classes were confined to special trains and cross-country workings while the V2s were usually ones which had been taken for a 'spin' by Annesley drivers. This meant that a crew would run an engine on a Marylebone train and return it the same day to Annesley having taken advantage of the fact that it had been rostered to work a turn from the Eastern Region to Annesley.

By the end of 1962 the situation was becoming clearer with the Great Central. Marylebone had closed to goods trains, Neasden was now derelict and Leicester shed was holding on by a thread, all of which aroused great concern in the village of Woodford Halse in Northamptonshire. This was a major depot on the Great Central and a dedicated railway village with no other form of industry. Even though the depot was flourishing, the future of the whole line was a subject of debate as the economic situation regarding British Railways as a whole grew worse. Such was the feeling that the Woodford Halse ASLEF Branch held a special meeting where a resolution was made which deplored the callous decision put forward by the then Minister of Transport, Ernest Marples. At this juncture Woodford had not yet felt the impact of any closures but the line as a whole was being killed off by being made unattractive to the travelling public with journey times of three hours from London to Nottingham, no refreshment facilities on stations or trains, poor time tabling and scheduling together with a general disruption to an otherwise efficient and profitable railway.

Suffice to say the line was being deliberately killed off, to ease the burden put on the coffers of British Railways. The first blow for Woodford came with the Beeching Report which suggested the closure of the station but made no mention of the depot. This would have effectively meant that those people without cars would have been isolated from the rest of the world, as Woodford is a long way from any major centre. In reality, during 1963 no action was taken to close Woodford Halse depot or the station, instead the focus was turned on the removal of the local services along the line as it was claimed that the intermediate stations did not provide sufficient revenue to justify their retention. To prove the point that the Great Central attracted

Leicester GC depot at 7.40am on 12th March 1961, three years before closure, with a full complement of motive power on shed. Outside the shed is BR Standard 5 73157 and B1s Nos61137 and 61380 while Leicester's sole diesel shunter hides away on a siding by the shedmaster's office, right. (K. C. H. Fairey)

insufficient revenue it was made the subject of a documentary film for the Beeching Report and when the line was filmed at 4 o'clock one afternoon, a V2 was shown drawing into the platform at Brackley Central with a long train and of course only a handful of passengers boarded the train. To use this as proof to shut stations or a line is biased because the pattern on any railway varies according to the time of day as to how busy it is. The fact that only one train was shown proves that the closure plan and the way it was carried out was an example of cold blooded, ill conceived action without any element of fairness. The Great Central was defenceless and in March 1963 on the 4th, stopping train services were withdrawn between Aylesbury and Rugby, Nottingham and Sheffield and on the Chesterfield loop. This meant the closure of the following stations. Some of them as shown in the table right still continued to function for freight traffic for a year or two longer.

Name of station	Passengers	Goods
Calvert	4.3.1963	4.5.1964
Finmere	4.3.1963	5.10.1964
Helmdon	4.3.1963	2.11.1964
Charwelton	4.3.1963	4.3.1963
Whetstone	4.3.1963	4.3.1963
Belgrave & Birstall	4.3.1963	–
Rothley	4.3.1963	4.3.1963
Quorn & Woodhouse	4.3.1963	4.3.1963
Ruddington	4.3.1963	–
Arkwright Street	4.3.1963*	–
Bulwell Common	4.3.1963	4.3.1963
Hucknall Central	4.3.1963	4.3.1963

* Reopened in 1967, closed again 1969.

The withdrawal of passenger trains meant that many little communities became isolated and the possession of a motor car became a necessity if one wanted to travel. Indeed if one wanted to board the train to London Marylebone, Leicester Central, Nottingham Victoria, Rugby etc one now had to travel to the nearest station that was open on the Great Central London Extension. Those that did remain open were Brackley Central, Woodford

A Cravens diesel multiple unit which has just arrived at Leicester Central seems dwarfed by the lavish platform and canopy designed for better things, such as the 'Sheffield Specials', 'Master Cutlers' and 'South Yorkshireman' expresses. The train, judging by the number of people standing on the 'down' platform on this occasion, seems to be enjoying a well patronised run to Nottingham.
(N. G. Steele)

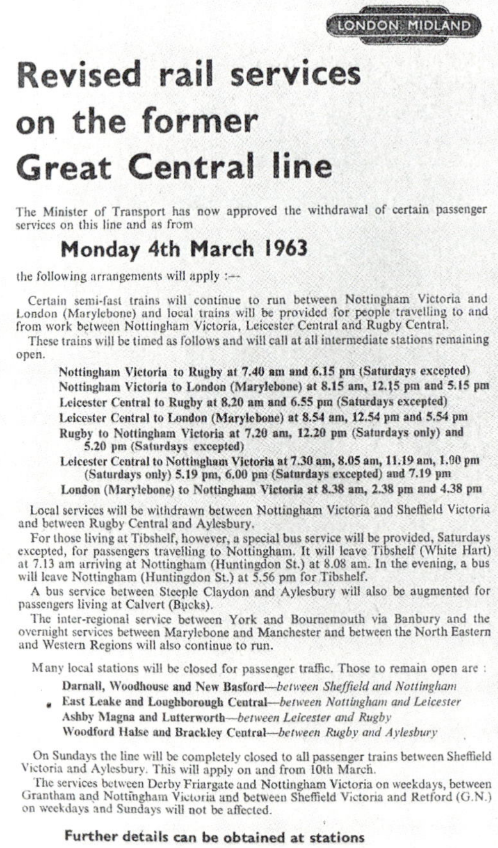

LONDON MIDLAND

Revised rail services on the former Great Central line

The Minister of Transport has now approved the withdrawal of certain passenger services on this line and as from

Monday 4th March 1963

the following arrangements will apply :—

Certain semi-fast trains will continue to run between Nottingham Victoria and London (Marylebone) and local trains will be provided for people travelling to and from work between Nottingham Victoria, Leicester Central and Rugby Central.

These trains will be timed as follows and will call at all intermediate stations remaining open.

Nottingham Victoria to Rugby at 7.40 am and 6.15 pm (Saturdays excepted)
Nottingham Victoria to London (Marylebone) at 8.15 am, 12.15 pm and 5.15 pm
Leicester Central to Rugby at 8.20 am and 6.55 pm (Saturdays excepted)
Leicester Central to London (Marylebone) at 8.54 am, 12.54 pm and 5.54 pm
Rugby to Nottingham Victoria at 7.20 am, 12.20 pm (Saturdays only) and 5.20 pm (Saturdays excepted)
Leicester Central to Nottingham Victoria at 7.30 am, 8.05 am, 11.19 am, 1.00 pm (Saturdays only) 5.19 pm, 6.00 pm (Saturdays excepted) and 7.19 pm
London (Marylebone) to Nottingham Victoria at 8.38 am, 2.38 pm and 4.38 pm

Local services will be withdrawn between Nottingham Victoria and Sheffield Victoria and between Rugby Central and Aylesbury.

For those living at Tibshelf, however, a special bus service will be provided, Saturdays excepted, for passengers travelling to Nottingham. It will leave Tibshelf (White Hart) at 7.13 am arriving at Nottingham (Huntingdon St.) at 8.08 am. In the evening, a bus will leave Nottingham (Huntingdon St.) at 5.56 pm for Tibshelf.

A bus service between Steeple Claydon and Aylesbury will also be augmented for passengers living at Calvert (Bucks).

The inter-regional service between York and Bournemouth via Banbury and the overnight services between Marylebone and Manchester and between the North Eastern and Western Regions will also continue to run.

Many local stations will be closed for passenger traffic. Those to remain open are :

Darnall, Woodhouse and New Basford—*between Sheffield and Nottingham*
East Leake and Loughborough Central—*between Nottingham and Leicester*
Ashby Magna and Lutterworth—*between Leicester and Rugby*
Woodford Halse and Brackley Central—*between Rugby and Aylesbury*

On Sundays the line will be completely closed to all passenger trains between Sheffield Victoria and Aylesbury. This will apply on and from 10th March.

The services between Derby Friargate and Nottingham Victoria on weekdays, between Grantham and Nottingham Victoria and between Sheffield Victoria and Retford (G.N.) on weekdays and Sundays will not be affected.

Further details can be obtained at stations

BR 35000 AD 470 February 1963

On 12th October 1963, GWR 4300 class 2–6–0 No.6368 and 2251 class 0–6–0 No.2246 are seen leaving Woodford Halse with the Locomotive Club of Great Britain's 'The Thames, Avon & Severn Rail Tour'. ***(M. D. Marston)***

Halse, Rugby Central, Lutterworth Ashby Magna, Leicester Central, Loughborough Central, East Leake, Nottingham Victoria and Basford. These stations were served by the semi-fast trains/cross-country workings and those between Rugby and Nottingham also by the remaining commuter service. The latter had made it necessary to effect changes in the running of the long distance trains with the result that the 12.40pm ex Marylebone now departed at 2.40pm while the 8.40am was timed to leave Nottingham nearly half an hour earlier. Not only were timings altered but also the coach formations with all trains having four carriages except for the 5.15pm which carried seven carriages. The first train of the day had in fact been converted to a diesel multiple unit working and this ran from Marylebone.

The next changes to occur on the Great Central were the removal of the Banbury to Woodford local service in June 1964. In the same year the shed at Leicester ceased to operate as its role had been taken away, namely to provide relief engines for long distance workings and freight trains, which diminished rapidly in the final few years. However during 1963–64 the Great Central saw an increase in the number of parcels trains and sleeping trains which were transferred to Marylebone or routed into the Great Central system over the Bletchley to Calvert junction line. Motive power was also varied and the locomotives which had appeared with 'London Midlandisation' in 1958 were now becoming mechanically worn out and their numbers dwindled, being replaced by Stanier and Standard Class 5 4–6–0s.

In the last two years the final rites were enacted by the closure plan so quickly that it was hard to believe the Great Central had ever been a flourishing railway. The first of two lightening stages to remove all freight traffic was brought into effect on 1st March 1965 with the diversion of all through goods traffic. It involved about 470 wagons daily, mostly going to Wales and the South West via the old

92030

61051

Left, top: A BR Class 9F 2–10–0 No.92030 normally used for freight proves itself by making light work of a twelve coach train forming the summer Saturdays only, Skegness – Leicester express, 8th August 1964. Here the train is seen rounding a curve near Belgrave & Birstall.
(M. D. Marston)

Left, bottom: After the closure of Leicester shed in 1964 engine exchange on long distance working at Leicester Central became increasingly difficult, as locomotives had to travel down light engine from Annesley to work trains north. Here, carrying light engine headcode Class B1 4–6–0 No.61051 heads for Leicester past Belgrave & Birstall. *(M. D. Marston)*

On 8th August 1964 BR Standard Class 5 No.73000 is seen hauling an interesting variety of coaches near Belgrave & Birstall with a relief holiday express. The first vehicle is a Great Western Hawksworth design while the second is of LMS design with the rest of the train comprised of British Railways' Southern Region stock. *(M. D. Marston)*

Occasionally diesels worked turns over the Great Central and in this picture English Electric Type 3 Co-Co diesel electric No.D6743 (later No.37043) is seen passing Swithland signal cabin with the Bournemouth to Newcastle train. Note the leading coach which is of Gresley design and the BR Mk1 buffet car, sixth vehicle in the train. *(M. D. Marston)*

An atmospheric shot of class 9F No.92030 with a mixed freight having just cleared the approaches to Dunton Bassett, heads north past Ashby Magna signal cabin and goods yard.
(M. D. Marston)

Stratford on Avon & Midland Junction Railway from Woodford Halse. The second was the biggest step towards the closure of the Great Central line and this involved the re-routing of mineral traffic. The latter meant the removal of some 720 wagons a day. This mainly comprised coal from the East Midlands and the South Yorkshire pits, which was diverted to the Birmingham – Barnt Green – Bristol route and that from Banbury to London. The new arrangements for the Great Central line mineral traffic added to the workload of Toton yard, but there was sufficient space to absorb the extra volume. The removal of freight from the Great Central on 12/13th June 1965 would not really have been totally dependent on a yard like Toton as yard processing was giving way to block train workings and greater concentration schemes together with easier operating practises. Under the latter conditions the Great Central yards at such places as Annesley and Woodford would have been run down and closed in any event. The closure of these yards meant the disappearance of the 'Runners' and 'Windcutters' which were described in the previous chapter.

The 12th/13th June, which was when the new timetable for 1965 came into effect, brought with it the total closure of Woodford Halse depot with disasterous consequences for the local inhabitants which will be discussed at a later stage. With the closure of the yards, so went the numerous signalboxes along the line which ensured the safe passage of the many freight trains which had run over the route. After the shed had shut the last six engines at Woodford were taken away and scrapped while the onus fell on Annesley depot to provide engines for the remaining services on the whole line. Fortunately relief was to be provided by the depot at Banbury which had been transferred to the London Midland Region from the Western Region, otherwise it would have been difficult to imagine how the line could have carried on.

The line was by now in such a state that it was almost impossible to run a service as trains broke down, were unreliable, ran out of water and would often get stuck between the incredibly long signalling block sections. Because of the nature of things it would be impossible to get a replacement engine. Most of the lineside equipment had been stripped, so it was not even possible to inform anybody that the train had broken down! To ease the problem of unreliability during the final part of 1965 some 'Britannia' class engines returned to Banbury though they did very little to improve the situation regarding the motive power. In December 1965 with the last major reshuffle of steam, the engines disappeared and in the same month Annesley shed handed over all its responsibilities for the provision of motive power to neighbouring Colwick depot in Nottinghamshire. Services lingered on for another year following the official declaration for

'Britannia' class 4-6-2 No.70052 **Firth of Tay** *arrives at Leicester Centrals 'down' platform with the 5.15pm semi-fast to Nottingham on 20th October 1965. The tracks on the outside of the down loop have already been removed as they were no longer needed as carriage or engine holding sidings. Far right is Leicester South signal cabin.* *(M. D. Marston)*

closure of the whole line. On 3rd September 1966 the Great Central ceased to operate as a trunk route, and its services were either transferred or withdrawn. These were the three semi-fasts in each direction which were discontinued, the York to Poole train which was re-routed, the Manchester to

LMSR class 5 4-6-0 No. 44847 eases out of Rugby Central on the 5.15pm ex Nottingham to Marylebone semi-fast on 14th July 1966. Note the signal box nameboard which displays the name Rugby Station rather than Rugby Central. *(K. C. H. Fairey)*

Left: On 15th July 1966 a very worn and scruffy looking Class 5, No.44847 with a train of Mk 1 carriages sits in the 'up' platform at Rugby Central, whilst taking a drink before heading off for Woodford Halse and Brackley Central, on its way to London, Marylebone. *(K. C. H. Fairey)*

Left, bottom: 1st September 1966 and all is slowly coming to an end after 67 years of operation. 'Black 5' No.44840 heads the 4.38pm semi-fast ex Marylebone out of Woodford Halse on its way to Rugby and Nottingham Victoria. The platform on the right became derelict two years previously with the end of services to Banbury, though the line through the platform continued to be used June 1965 when goods services ceased. *(K. C. H. Fairey)*

Marylebone night train which was withdrawn and the York to Bristol and Sheffield to Swindon mail trains which were also re-routed. Added to these were two newspaper trains from Marylebone to Nottingham at 1.40am and 3.40am and a parcels train from Nottingham to London which were all discontinued.

On the last day services were the same, save for the strengthening of trains with extra coaches and the running of one special from Waterloo to Sheffield and back. Many people turned out to witness the end of the 'Last Main Line' to London and some of those who witnessed the closure had been present at the opening, proving how short the life of the route had been. Although main line services had ceased the section between Rugby and

A Derby built DMU calls at a rather derelict and dreary Ashby Magna station in 1968 on its way up to Lutterworth and Rugby Central from Nottingham Arkwright Street. *(N. G. Steele)*

Class 5 4–6–0 No.44858, one of the last steam locomotives on the London Extension is seen on the 4.38pm ex Marylebone to Nottingham Victoria semi-fast, passing Marylebone yard cabin on 18th August 1966. This was 16 days before closure of the line as a trunk route. *(K. C. H. Fairey)*

Weekday Cross junction which lay to the south of Nottingham Victoria as seen from the window of a diesel multiple unit in May 1967. The line to the left was used by Grantham trains until July 1967 when they were diverted to run into Nottingham Midland station. Of interest is the precariously built signal box incorporated into the viaduct structure. ***(N. G. Steele)***

Train Service

Nottingham Arkwright Street and Rugby Central.

1 DECEMBER 1968 until Further Notice.

					SO			SX
NOTTINGHAM Arkwright St.	dep.	07 50	08 22	12 27	13 55	16 17	17 34	18 52
EAST LEAKE	dep.	08 03	08 35	12 40	14 08	16 30	17 47	19 05
LOUGHBOROUGH Central	arr.	08 10	08 42	12 47	14 15	16 37	17 54	19 12
„ „	dep.	08 11	08 43	12 48	14 15	16 33	17 55	19 13
LEICESTER Central ...	arr.	08 24	08 56	13 01	14 29	16 51	18 08	19 26
„ „ ...	dep.	08 26	08 58	13 03	14 31	16 53	18 10	19 28
ASHBY MAGNA ...	dep.	08 41	09 13	13 18	14 46	17 08	18 25	19 43
LUTTERWORTH ...	dep.	08 48	09 20	13 25	14 53	17 15	18 32	19 50
RUGBY Central ...	arr.	08 57	09 29	13 34	15 02	17 24	18 41	19 59

					SO			SX	
RUGBY Central ...	dep.	—	07 10	10 30	12 30	15 05	16 20	17 37	18 55
LUTTERWORTH ...	dep.	—	07 19	10 39	12 39	15 14	16 29	17 46	19 04
ASHBY MAGNA ...	dep.	—	07 27	10 47	12 47	15 22	16 37	17 54	19 12
LEICESTER Central ...	arr.	—	07 40	11 00	13 00	15 35	16 50	18 07	19 25
„ „ ...	dep.	07 10	07 43	11 02	13 05	15 37	16 55	18 12	19 30
LOUGHBOROUGH Central	arr.	07 21	07 54	11 13	13 16	15 48	17 06	18 23	19 41
„ „	dep.	07 22	07 55	11 14	13 17	15 49	17 07	18 24	19 42
EAST LEAKE	dep.	07 30	08 03	11 22	13 25	15 57	17 15	18 32	19 50
NOTTINGHAM Arkwright St.	arr.	07 42	08 15	11 34	13 37	16 09	17 27	18 44	20 02

Notes : SO—Saturdays only. SX—Saturdays excepted.

This service will provide SECOND CLASS accommodation only.

Passengers will be able to obtain tickets, **between stations served by this Service only,** from the Guard in charge of the train.

Accommodation will be provided for the conveyance of cycles, perambulators, etc., accompanied by passengers, who will be responsible for the removal of these articles from the stations.

Unaccompanied traffic will not be conveyed.

Season tickets, **between stations served by the Services only,** will be issued at Nottingham Midland, Leicester London Road and Rugby Midland Stations.

	Nottingham		East Leake		Loughborough		Leicester Central		Ashby Magna		Lutterworth		Rugby Central	
To:	S	R	S	R	S	R	S	R	S	R	S	R	S	R
Nottingham ...	–	–	2/6	3/9	3/9	5/–	6/3	7/6	8/9	11/–	9/9	13/6	11/9	16/–
East Leake	2/6	3/9	–	–	1/4	2/6	4/3	5/6	6/6	10/–	7/9	12/–	9/6	14/6
Loughboro Central	3/9	5/–	1/4	2/6	–	–	2/9	4/6	5/6	9/–	6/3	11/3	8/3	14/3
Leicester Central ...	6/3	7/6	4/3	5/6	2/9	4/6	–	–	2/9	4/9	4/–	6/3	5/6	9/–
Ashby Magna ...	8/9	11/–	6/6	10/–	5/6	9/–	2/9	4/9	–	–	1/2	2/3	3/–	5/6
Lutterworth ...	9/9	13/6	7/9	12/–	6/3	11/3	4/–	6/3	1/2	2/3	–	–	2/–	3/9
Rugby Central ...	11/9	16/–	9/6	14/6	8/3	14/3	5/6	9/–	3/–	5/6	2/–	3/9	–	–

The return fare quoted above is that for Cheap Day Return.

British Rail

Issued by British Railways
Divisional Manager, Furlong House,
Middle Furlong Road, Nottingham.

BR 3500 **AD184** December, 1968

Nottingham was retained for a commuter service which ran to the timetable below. The latter was run by a series of diesel multiple units and was a far cry from the trains which had run in the heyday. The stations on this short stub at Rugby, Lutterworth, Ashby Magna, Leicester, Loughborough, East Leake and Nottingham Victoria had been reduced to the status of unstaffed halts. It was not long before changes were made even to this little commuter service. Nottingham Victoria was now far too large to cope with just one two-car diesel multiple unit disturbing the tranquillity of its vast concourse three or four times a day, so it was decided to give the station at Arkwright Street a facelift, and this was to be the terminus for the final two years.

There were two reasons for the closure of Victoria, one of which could also be the reason for it remaining open after the withdrawal of through services to London. The latter was due to the fact that Victoria still served Grantham trains after closure as a trunk line station. This service like the one to Rugby was also worked by diesel multiple units. A new link at Netherfield was provided to allow the Grantham trains to run in to the Midland station at Nottingham which opened on 3rd July 1967. The second reason was that the site was

The impressive main building of Nottingham Victoria station and the clock tower in August 1967, month before closure and almost immediate demolition to make way for what is now known as the Victoria Centre. This is a shopping complex with all the 'mod cons' for its customers but the railway element has refused to die as the clock tower still remains as a monument to the Great Central. ***(N.G. Steele)***

wanted for a major development to be known as the Victoria Centre. Even though the Grantham services had been transferred away successfully, the development was still impeded by the delay in preparing Arkwright Street as the terminus and in fact Victoria was not closed until September 1967. After this date demolition of the grand old building began with rapidity and soon there was nothing left except for a big hole with two tracks in the centre of it, which were retained for freight movements between Bagthorpe Junction and Weekday Cross Junction. The two tracks mentioned above lasted only a year for freight trains and were finally removed from service on the 25th May 1968 after a drastic reorganisation of railways in the Nottingham area when the traffic from the old LNER lines was diverted to those of the Midland.

Almost a year later even the passenger service to Rugby was to be scrapped. It came as no surprise as it had been forecast for years previously and

A sad day for Nottingham Victoria as the last diesel multiple unit arrival from Rugby Central draws into platform 9 on 4th September 1967. After this passenger trains terminated at Arkwright Street station to the south of Weekday Cross. ***(N.G. Steele)***

The impressive façade of the booking office at Rugby Central in 1968, then the terminus of the scant service which plied to and from Nottingham Arkwright Street. The writing is on the wall though, even for this service as the posters by the telephone kiosk advertise the withdrawal of passenger services, a sad day for the London Extension. *(N.G. Steele)*

A busy scene viewed from the cab window of a Cravens 2-car diesel multiple unit as it waits for customers in the 'down' platform at Rugby Central on the last day of operations, 5th May 1969. In the mist in the background are the signal cabin, water tower and entrance to the goods yard. Just south of the bridge in Clifton upon Dunsmore cutting, the line ended at a pair of buffer stops from which point the old main line to Aylesbury and Banbury had been lifted. *(N.G. Steele)*

indeed it would have been hard to imagine a line like it surviving as it effectively ran from nowhere to nowhere. It had no services which it might give revenue to or take revenue from and was run in a rather haphazard fashion to discourage passengers from using it. After the last train ran on 5th May 1969 villages like Ashby Magna, East Leake and Lutterworth had to rely on road transport while in Loughborough, Rugby Leicester and Nottingham little hardship was felt as there was still another rail facility in these towns.

By the time the closure plan for the Northern section of the old Great Central had been instigated, track lifting between Calvert and Rugby was almost complete and this included the section between Banbury and Woodford. Work had also been pro-

Right, top: Lutterworth, the next stop after Rugby on the day of closure to all services. This was one of the few stations which remained open until the very end, albeit in a very sorry state of repair as is evident from the condition of the glass canopy over the stairway. At the far end of the platform a handful of enthusiasts gather to see the last departure ever from the station, destined for Rugby Central. *(N.G. Steele)*

Right: Another view of Lutterworth at closure, from the old goods yard showing the island platform in its entirety. A couple of features here are worthy of note. The gentlemen's lavatory appears to have had the original ornamental capping removed and a new row of brickwork added around the top of the walls, a cycle shed has been added, and also note the extra building at the other end of the platform which has been isolated from the others by the stair well. *(N.G. Steele)*

LUTTERWORTH

It's the end of the line for an impermanent way

● It was the end of the line for the last part of the Nottingham - Manchester section of the old Great Central track today.

British Rail workmen are seen dismantling and hoisting away the metals on an embankment on the northern boundary of Hucknall.

60FT. SECTIONS

The gang lift the permanent way in 60ft. sections in an operation which has already removed the "up" line from Pilsley and is being pushed forward to New Basford.

Soon, the only surviving section of the old Great Central line between Nottingham and Rugby will disappear under British Rail's plans for the future national railway network.

gressing on the section between Bagthorpe and Annesley as one of the newspaper cuttings shows. The actual removal of the railway had all kinds of effects on people. For those who travelled on it there was the need to find alternative means of transport which may have meant moving away from the communities which established themselves around the Great Central. Many relied on the line to take them to work and to give them a chance to go on days out. For those who worked on the line, closure meant loss of jobs and depending on which area one looked at depended on the feelings of the staff and their families. At Annesley and Nottingham whilst some 1,000 people were made redundant it was an area of great industry and there were also other railway systems where employment could be found, and so, little upheaval was caused. The situation at Loughborough was similar as there was alternative employment both within the railway and in engineering.

Those men who had worked the stations in between now found themselves travelling to work

RAIL CUT PLANS

CLOSURE of the Great Central line would cause hardship to passengers, the East Midland Transport Users' Consultative Committee has told the Minister of Transport.

Possible means of alleviating possible hardship have been suggested by the committee to the Minister, but these have not been revealed.

The Minister's final decision will be made later.

Proposals

The British Railways Board proposes to withdraw all passenger services between Sheffield Victoria and Aylesbury, and between Woodford Halse and Banbury — except between Sheffield Victoria and Woodhouse, and between Nottingham (Arkwright - street) and Rugby Central—and to close Nottingham Victoria, Woodford Halse and Brackley Central stations.

The plans have been considered by the committee, and objections made to the proposals in writing, and at the public hearing at Leicester in October, have also been considered.

In their report to the Minister, the committee said the withdrawal of the service would inflict hardship on various categories of passengers, using the through services from the North to the South and South-west.

BRITISH RAILWAYS BOARD

PUBLIC NOTICE — TRANSPORT ACT, 1962

WITHDRAWAL OF RAILWAY PASSENGER SERVICES

The London Midland Region of British Railways hereby give notice in accordance with Section 56 (7) of the Transport Act, 1962, that they propose to discontinue all railway passenger services on the former Great Central line between Nottingham Victoria and Rugby Central and to close the following stations :—

Nottingham (Arkwright Street)
East Leake
Loughborough Central
Leicester Central
Ashby Magna
Lutterworth
Rugby Central

Note : The Rugby - Nottingham rail passenger services which now operate to and from Nottingham Victoria will from 4.9.67 operate to and from Arkwright Street Station at Nottingham.
The Minister of Transport has already sanctioned this alteration, in giving her consent to the Board's proposal to close Nottingham Victoria Station.
It appears to the Board that the following services will be available :—

Existing Services by Rail :	Nottingham Midland-Loughborough Midland Nottingham Midland-Leicester London Road Loughborough Midland-Leicester London Road
Existing Services by Road :	
Provided by South Notts. Bus Co. Ltd.	Nottingham - East Leake Nottingham - Loughborough East Leake - Loughborough
Provided by Trent Motor Traction Co.	Nottingham - Loughborough Loughborough - Leicester
Provided by Barton Transport Ltd.	Nottingham - Loughborough Nottingham - Leicester
Provided by Birmingham and Midland Motor Omnibus Co. Ltd.	Leicester - Ashby Magna Leicester - Lutterworth Leicester- Rugby Lutterworth - Ashby Magna Rugby - Ashby Magna Lutterworth - Rugby

Additional Services Proposed by Road.
Subject to the approval of the Traffic Commissioners, alternative Bus Services on Monday to Fridays will be operated by the Birmingham and Midland Motor Omnibus Co as under:—

Approx. 06.50 from Rugby via Lutterworth and Ashby Magna Turn, arriving Leicester at approx. 07.50.
" 08.00 from Leicester via Ashby Magna Turn and Lutterworth, arriving Rugby at approx. 08.50.
" 10.00 from Rugby via Lutterworth and Ashby Magna Turn, arriving Leicester at approx. 10.50.
" 13.00 from Leicester via Ashby Magna Turn and Lutterworth, arriving Rugby at approx. 13.50.
" 18.10 from Leicester, arriving Ashby Magna at approx. 18.45.

Any user of the rail service which it is proposed to discontinue and anybody representing such users may lodge an objection in writing within six weeks of August 4th, 1967, i.e. not later than 15th September, 1967, addressing the objection to :—

The Secretary,
Transport Users' Consultative Committee for the East Midland Area,
44 Friargate, Derby.

If any such objection is lodged, the services cannot be discontinued until the Transport Users' Consultative Committee has considered the objections and reported to the Minister of Transport, and the Minister has given her consent to the closure under Section 56 (8) of the Transport Act, 1962.

The Committee may hold a meeting to hear objections. Such a meeting will be held in public and any persons who have lodged an objection in writing may also make oral representations to the Committee.

If no objections are lodged to the proposal the service will be discontinued on 18th September, 1967.

by bus or car if they were lucky enough to have one, to what would probably have been a totally different environment to that which they had been used to on the Great Central. The bus services which replaced the Great Central trains are shown on the 'Withdrawal of Railway Passenger Services' notice seen here which was a regular feature on the stations between Rugby and Nottingham in the last few years. The press cutting shows the original reaction to the announcement that the semi-fast services were to be withdrawn and shows the British Railways Board proposals. When considering closures the question of economy is always put before the needs of the passenger and this is made clear by the report

City-Rugby line to end

MR. RICHARD MARSH, Minister of Transport, has rejected reuests by East Midlanders for the continuance of the former Great Central Line between Nottingham Arkwright Street and Rugby Central.

Among MPs and local organisations who have compaigned to keep open this line, Mr. Tony Gardner, MP for Rushcliffe, today received an official statement from the Ministry of Transport saying that the Minister had made his decision after studying reports from the East Midlands Transport Users' Consultative Committee and the East Midlands Economic Planning Council.

The closure of this route will involve discontinuance of all railway passenger servies between Arkwright Street and Rugby. The Minister felt unable to ask the Railways Board to continue this service in spite of the hardship to passengers who now use the line.

Additional bus services are being provided on the direct route between Rugby and Leicester and buses will be routed by the Leicester-London road to provide connections with trains on the main line serving Loughborough and Nottingham.

'Ghost' of the old GC

FROM next Monday, Nottingham Victoria Station—one of the most imposing pieces of railway architecture in the provinces — will be little more than a "ghost" station, entirely unmanned except for two car park attendants in the yard, and signal staff.

This is when the Nottingham-Grantham diesel service is switched to the Midland Station.

Climax

The only other passenger trains using Victoria go to and from Rugby.

"This is the climax to a sequence of events bringing an end to the Great Central line," commented Mr. Roy Trench, Nottingham area official of the NUR.

"It is regrettable, but every effort has been made by the union to keep the line open."

'City Line to End'. It is without doubt that the removal of the line caused great hardship to passengers and staff alike. There is also another aspect to consider and that is the demolition aroused an untold feeling of nostalgia and sadness for the people of Nottingham and Leicester alike as the cuttings show by their headings. One must not forget that the Great Central had been a part of every day life for many. Even today it is hard to forget the line as traces of it can still be found in Nottingham and Leicester in the form of viaducts, bridges, tunnels, buildings and, in the case of Nottingham, the clock tower still stands defiantly and proudly in memory of the station to which it once heralded the entrance. In Leicester the station approach and parcels depot still remain as a reminder though the actual station has been covered with industrial units. For Leicester the extract from Mr Leicester's Diary of 5th May 1969 sums up the feelings exactly which one man had about the Great Central, and also it gives a true

The approach to Leicester Central station in the last days of the London Extension. Despite the demolition of the station itself, the building in the background has survived as has the lintel (hidden behind the building on the right) which proclaims the entrance to the parcels office in a rather ornate style.
(N.G. Steele)

MONDAY, MAY 5, 1969

Mr. Leicester's DIARY

THE last train had gone and Mr. Albert Fawkes locked up his signal-box on Saturday evening, picking his way along the track the few yards to the deserted platforms of Leicester's Central Station.

He secured the empty waiting rooms and then finally locked up the station for the last time. And an important era in Leicester rail history closed too.

The last surviving stretch of the Great Central Line—Nottingham - Leicester - Rugby—over which the axe had long wavered, had finally yielded to economics and the profit and loss account.

The station, ingeniously constructed on a long snaking viaduct, had been "unmanned" for some time, the job of opening up in the morning and locking up at night falling to signalman Fawkes or his alternate shift signalman Mr. Douglas Brown.

Mr. Fawkes, 34 years a local signalman, has spent the last nine at Central Station and witnessed its rapid decline.

"We were exceptionally busy when I started here. I just can't believe it's closing, seeing how the position was in 1960.

"There were 90 or 100 trains through here per eight-hour shift. Five signal boxes were in operation. You never got a chance to sit down. But the traffic has gradually been taken over the past three or four years."

On Saturday, from the single working signal-box, he supervised the last of the handful of commuter trains which latterly had passed through the station on weekdays.

As any life-long railman should be, Mr Fawkes was sad to see the end of the station where the rust and weeds had already started to encroach on its disused sections.

"It takes a bit of believing when you have spent the whole of your life on the iron road. You just cannot agree with it," he said.

Mr Fawkes, who lives at Bonville Place, Leicester, hopes to carry on as a top signalman elsewhere in the city.

After locking up, he went home in his car – ironically the vehicle which is responsible in many ways for his having to perform the last act.

'JUST A PLACE ON THE MAP'

100 angry women in rail protest

"THEY think we are just a place on the map, not people," a woman said at a meeting organised by women's organisations at Woodford Halse last night to protest against proposed rail closures which threaten the village. She said no-one from the Ministry ever came to see places or to hear people's views.

Nearly 100 women attended the meeting and unanimously passed a resolution protesting at the proposed closure of the Great Central line, and the Woodford Halse depot, and pointing out the hardships, worries and inconveniences it would cause.

The only men present were members of the deputation appointed to see the Under Secretary to the Minister of Transport to protest at the proposed closures, and the leader of the deputation, Mr. Ivor Wilde, urged the women to arm them with as many facts as they could to strengthen their case at the Ministry talks.

One woman said if they were going to keep the railway open it should be a permanent job; it was no good going on for just a few years. When men got to 40 years of age it was not easy to get another job.

Old relatives

A spokesman from Weston W.I. said that if the line was closed they would have difficulty in travelling northwards because the train was the only service to Rugby. Railwaymen woud have to dig up their roots and families would have to leave their old relatives behind. Who would look after them ?

Another woman pointed out that youngsters who attended evening classes would have no means of getting home which was a "frightening" thought. Hospital visiting would also be a problem, the train being the best connection with Banbury.

A railwayman, Mr. W. J. Chalmers, said the G.C. line had been "sabotaged" by Midland Region. They used to provide wide and first class services until they were interfered with.

"I can't move my home and I don't want to leave the railway," he said, "but I certainly don't want to move to a town and rush here and rush there. I like it here—most men do.

Mother's query

A mother asked who would look after the children if her husband had to go away to work and she was taken ill.

Wives of older men expressed their fears that their husbands would not get the lodging allowance and would certainly not be able to get other work.

They pointed out that most were buying their own homes and would not be able to sell them in a dying village. They would not be able to afford the high prices of the towns or find their own homes there.

One pointed out that other rail centres were also being shut down and asked where the men would find other railway work.

Lodging allowance

"They say if we have to work away from home they will pay £4 a week lodging allowance but that will not go far in London or Crewe.

"We can't go to Northampton because there is talk that they are going to close that in 1966. If you go for another job you are told you are too old. If they are going to close the railway, what we would like to see is a factory built, before they close it, to provide alternative employment."

Mr. F. Sanders said the closure would hit the traders badly. They would have difficulty in getting supplies—papers, coal and goods—and with people leaving the village, trade would decline.

The chairman of the meeting, Mrs. F. Brown, president of the W.I. said it was no good the women putting their heads in the sand as the men had done for so long. They must do something about it.

Another point of view put forward was a reminder that in the bad winters, the train was the only means of getting food to the village. Woodford was only on minor roads and they were blocked in bad weather.

Mr. A. C. Jordan said that railwaymen were told that their free travel passes and reduced fares were part of their wages but if there were no trains to use them on, they were being deprived of part of their pay.

"I can remember being marched from school to see the first train to come into Woodford," he added. "I certainly don't want to see the last."

Mr. S. Allen, chairman of the parish council, said if the proposals went through, Woodford would be moribund. The council had been trying to get some light industry for three years and were still negotiating, but rather unsuccessfully.

account of how the Great Central was rundown and closed.

It often seems hard to imagine why something like a railway attracts such a following after closure, but I feel that it is due to the fact that in the decade when the Central closed it was regarded as something that would never die and always be functional and when it did disappear, the shock more than anything else prompted the need to immortalise it. The most powerful influence behind the latter was definitely the memories which the line generated as people watched the line being torn apart. There is no greater proof of this feeling than the huge piece written about Victoria in Nottingham as it was being demolished.

The piece conjures up all the former glory of the old station giving a little bit about its history and the way it ran and what services it provided etc, and evokes a real nostalgic effect. The feelings were such for keeping this lifeless line alive that numerous

DAY STATION DIED QUIETLY

No fanfares as Victoria is put to sleep

By Post and News reporter

THE end finally came to Nottingham Victoria Station on Saturday—not in a blaze of glory, but quietly and with a deathly hush.

The last act of the drama was played out in silence with nobody to applaud the end of a magnificent tradition.

Nobody was there to recall the past glories of the continually changing scenes of over half a century in the life of Nottingham.

Turn back the clock to the summer Saturdays of bygone years. Crowds throng the platforms beneath the lofty roof. A promising sun pierces the smoke and steam to spotlight the jostle of cases and faces.

Excited children clutching buckets and spades were marshalled into coaches for Skegness, Mablethorpe and Bournemouth for their once a year train ride to get away from it all.

CHANGE

A change in scene. A chill autumn night 50 years ago. Grey smoke and a haze of drizzle provide the back-cloth this time for khaki uniforms, kitbags and rifles.

There are brave smiles as the trains draw out, but the lace-edged handkerchiefs are damp clutched tightly in the gloved hands of the womenfolk. They wave to the disappearing tail-light on the last carriage. It is war and they know their men may not return.

The stage changes again. It is 1959 and a stream of trains arrived with thousands of cheering passengers. It was great news. Nottingham Forest had won the FA Cup and there was talk of nothing else as the crowds jostled over the footbridge.

Now only the theatre is left. The actors have gone. Victoria is officially dead, although not quite buried.

GHOST

The infrequent service to Rugby will carry on for a while until Arkwright-street station is brought back into use, and ghost-like, impersonal coal trains will rumble through.

But the few remaining staff have moved on to other jobs and soon the vast building on its 13-acre site will disappear.

Victoria was not an old station, but during its 67-year life it earned an affectionate place in the hearts of Nottingham people.

REAL

For thousands of people it was the first view of Nottingham and its great size never failed to impress them. For others the dark tunnels at each end symbolised departure into the unknown fortunes of a new life in the outside world.

To others it meant a daily journey to work, a long awaited reunion, or just a trip to the East Coast. Whatever you felt about the Victoria, it could not be ignored.

By virtue of its sheer size it always seemed to be a "real" railway station to local people and its square tower and green dome represented only one thing—travel.

Victoria opened in May 1900 by the Great Central Railway, was the most imposing station built on the main line then being constructed between Sheffield and London. It was shared by the Great Northern Railway, who served Derby, Grantham and the East Coast.

TOO LATE

The station was part of a magnificent dream which never came true. The idea of the promoters of the line was to make a through route from the Midlands and industrial North to Europe through a Channel tunnel.

Ironically the recent plan to drive a tunnel to France came too late to save the one line which was built with that very intention in mind.

The Great Central lived up to its name in every respect. Its services were fast, comfortable and highly competitive with other routes to the North. Its only failing was that it came too late, missing the full flood of the railway boom of the late 1890s.

It never made money and the trials of the First World War sealed its fate.

But its management thought big and Victoria Station, Nottingham, was a monument of their breadth of view.

PARADISE

Although the last passenger trains to run have been rail cars, the Board of Directors would not have been surprised. They were running a petrol electric service from the city at off peak times as long ago as 1912.

Until nationalisation in 1947 Victoria was a schoolboy's paradise of big green engines and varnished teak coaches, both under the Great Central and the LNER which took it over in the 1920s.

Until recent years famous trains such as the Master Cutler to Sheffield and the Yorkshireman called at Nottingham.

Now all is silent. Pigeons roost in the latticework of the girdered roof and the once shiny metals grow rusty and overgrown. Grime covers the windows of the waiting rooms and refreshment buffet and a pall of dust settles on the bench seats.

SUSPENSE

It is an unearthly atmosphere, a kind of suspense. The great building waits for the moment when it will tremble again—not with the rush of a thousand footsteps or the arrival of the trains—but the blows of the demolition worker's hammer.

But the Great Central will be remembered to the end. In clear letters over a booking office window, recently uncovered when a bookstall was removed from the main concourse, is written: "Great Central" with either side the words "In" and "Out."

Victoria Station will make way for what will become the city's largest office block.

The Board of Directors would have liked that. The ywould have understood the march of progress and borne the loss of their great station with fortitude.

They were proud of the motto emblazoned on the Great Central Coat of Arms—"Forward."

organisations both for locals and enthusiasts were set up to preserve some vestiges, one of these organisations forming the basis of the second part of this book, namely the Great Central Railway 1976 Limited, née Main Line Preservation Group.

So far we have considered the attitudes and feelings of the people who lived and worked on the northern half of the line and now we turn to the town which suffered most as a result of the closure of the Great Central and this was Woodford Halse. This was a dedicated railway settlement and much work was provided at the depot in some form or other. Some 650 men were employed prior to closure in the yards, carriage and wagon shop and locomotive shed which stabled 60 locomotives. Of these 200 went to other areas to find work on the railways or retired from the industry, while 450 were made redundant. Because of its location and the fact that it was only a small settlement there was no other industry other than farming, which the men could do locally. So, in order to get work it meant travelling to Daventry, Banbury or even further afield and more often than not the work would be degrading for skilled engine crews who found themselves in menial occupations.

Woodford did not let its railway go without a fight as the reports show, but it was more or less a foregone conclusion that the line would end up trackless, right from the first reorganisation of services and the mention of closure. The residents were helpless because they were up against a force that could make a good case for closure simply by disrupting the line's running practises and starving it

Right: The redundant south bay platforms of Leicester that once catered for trains running to Rugby and Woodford which called at all stations en route, are fading into oblivion. After a period of two years dereliction, the weeds begin to take over.
(John Clarke Album)

Right, bottom: Nottingham Victoria after the last departure for Rugby Central. The only traffic that passed through the station at this stage was freight trains which used the outer roads on the 'up' side until 25th May 1968, by which time this splendid station had been razed to the ground.
(John Clarke Album)

Weekday Cross junction after the lifting of the track into Nottingham Victoria and the singling of the line on the Grantham line and that to Arkwright street. The remaining line served as a headshunt which ran into the tunnel. Once trains were on the old main line they were propelled to East Leake and Rushcliffe to the Gypsum works and Ruddington when a new chord connected the line south of Ruddington to the Midland at Loughborough. Note the pipeline being constructed through the tunnel which was responsible for heating the Victoria Centre.
(John Clarke Album)

of its lifeblood by re-routing most of its traffic so it became a loss making entity. Most other authors have given their own feelings to the latter action and in some ways I believe it does not always show the actual views held by the people at the time when these events happened. Thus the only way I have been able to compile this chapter, so that it brings out the full sense of outrage and hardship felt at the time, has been to include a series of newspaper cuttings from the local paper for the Woodford area during the last six years. After that time Woodford lay derelict as the demolition men slowly cleared the site of buildings, track and other equipment. Thousands of tons of ballast and raw materials were reclaimed and sold off which had once formed the base of the marshalling yards that had been described as the 'Pride of Europe'. All that was left was a vast expanse of land with the odd building remaining with broken glass and bricks from the sheds and station here and there. Once the railway had been cleared the village felt the brunt of the effects it brought with it. People moved to find new jobs where there was better transport and facilities and many houses remained empty as Woodford no longer had any means of attracting people. Twenty five railway workers remained along with a few others but the situation was grim and has only just begun to pick up now, some 21 years after closure. For the few railway workers who remained, there was still the old railway club where they could go and reminisce over old times on the footplate while some used to walk over the old trackbed around the yards and station area. At Woodford, like the Loughborough area, a Great Central Railway Enthusiasts Association exists which shows films and slides of the line in its heyday.

The sad remains of the north end of Nottingham Victoria as seen from platform 12, formerly used for Derby trains and parcel trains. The track was lifted shortly after withdrawal of local services leaving the Derby trains to depart from platform 11 until their demise in 1964. Nothing remains now save for memories.
(John Clarke Album)

Part 2
The Great Central London Extension Today

GWR 5205 class 2–8–0T No.5224 having run round its train at Rothley eases gently under the road overbridge, before coupling up to its train for the return to Loughborough. *(Author)*

Chapter 5
Great Central Renaissance

Even though the final section of the Great Central Railway between Nottingham and Rugby had been closed to trains, there was a significant interest created in the line in its last years, concerning the possibility of retaining a section for preservation. It was with this in mind that a group of enthusiasts with the same determination and dream as Sir Edward Watkin – namely to acquire their own railway and run it, met in Leicester and the result of this meeting was the formation of the Main Line Preservation Group. Their specific aim was centred on preserving a suitable length of double tracked main line which remained in situ between Leicester, Abbey Lane and Nottingham, where trains representing the 'Big Four' pre-Nationalisation companies would be able to pass each other at speed. This section was chosen not only because of the outstandingly beautiful scenery through which it ran ie Swithland, the Charnwood Forest and Buddon Hill but also because it was a section which contained very few major engineering works.

The project like that of the original Great Central was to say the least very ambitious, but this did not deter the Main Line Preservation Group who pressed on to lease the $3\frac{3}{4}$ acre site of Loughborough Central Station which became the centre of operations. Prior to and during the leasing of the station buildings, the Group made all out efforts to publicise the new plans to rebuild part of the Great Central by holding exhibitions, film shows and displays which were presided over by Lord Lanesborough, the President of the Group and also a man whose lands the railway ran through. The response

Loughborough station entrance from Great Central Road, 3rd May 1980. *(P. D. Nicholson)*

generated by the exhibitions was very encouraging indeed and as a result work soon began on additional buildings at Loughborough Central.

Whilst negotiations for purchase of the line were in hand, as can be seen from the chart of the Group's progress below, a plan was also being put forward to incorporate part of the old line formation into a road scheme. Feelings ran so high, both from the locals and the Main line Preservation Group that eventually the scheme was to be abandoned.

most northerly point at Ruddington, making 18 miles in all had proved to be far in excess of what could be afforded in any realistic terms. British Rail was eager to dispose of these redundant assets, though at a price, and the fund raising by the initial appeal only managed to draw in some £50,000. This was in seven year covenants but it was not enough to satisfy British Rail who, whilst the time went by and the Trust's position was far from a strong one, imposed an extra £1,000 a month to prevent the track from being lifted. Around this time much progress was being made at Loughborough with the station buildings, while some small locomotives were beginning to appear on the scene. At last the possibility of operating a train service was becoming more and more of a reality, and in January 1973 Loughborough Station was opened to the public for the first time since its closure some five years before. For the benefit of the first visitors the operation was

Main Line Preservation Group Progress Chart

1969	
January	Main Line Preservation Group formed.
April	Survey of the line by the Group's land agent.
May	British Rail service ends at Loughborough. Line Operations Dept. set up. Station action groups set up to stop vandalism.
September	Talks with British Rail London Midland Estates about possible purchase of the line.
1970	
January	Top floor at Loughborough Central leased and renovation started to former Great Central state.
March	Proposition put to British Rail to buy the line.
July	Loughborough Town Council abolish plans to try to turn a section of line into a road between the A6 and A60.
August	Negotiations with locomotive owners complete subject to finalisation of details. Full information supplied to the Ministry of Transport for obtaining a Light Railway Order to enable operation.
1971	
January	Proposed new company, Main Line Steam Trust Ltd accepted by Inland Revenue and Charities Commission and given charitable status. Visit by the Railway Inspectorate giving general agreement for the scheme and satisfaction at the state of the line.
March	Leicester County Council abandon any interest in the road scheme for the Loughborough area.
June	Commencement of fund raising by Messrs John Rich and Co.
September	Main Line Steam Trust incorporated as taking over from the Main Line Preservation Group.
October	Commencement of main fund raising campaign.

So the project was now well under way with the Main Line Steam Trust having taken over the founding group and the battle was on to raise some £383,000. This was required to secure the purchase of the land between Loughborough and Birstall together with the double track between Loughborough and Quorn, and a single track between Quorn and Belgrave & Birstall.

The aims of the original group to run to the

limited to footplate rides, although this was soon to change with the introduction of the first regular runs for passengers to Quorn & Woodhouse. This operation was carried out under British Rail supervision and commenced on 30th September 1973. The engines that performed this service were No.377 *King Haakon 7, Littleton No 5, No.4, Robert Nelson, No.39* and *Marston Thompson and Evershed No.3*. These are detailed in Chapter 6.

Following expansion into the yard area by the Main Line Steam Trust, the old Great Central main lines were designated for a different purpose, namely the storage of rolling stock and locomotives. On the left is the former 'down' main line stabling the Norwegian State Railway's 'Conductorvogn' No.547 built by Strommens Vaersted in 1912 and class 21C No.377 Engine* King Haakon 7. *The ex 'up' line boasts a greater variety with a line up of tank engines and goods wagons with* Marston Thomson & Evershed No.3, *No.39,* Robert Nelson, *No.4* Hilda *and an Esso tank wagon. ***(M. Hooper-Immins)***

Loughborough in February 1973 and performed a fare share of the initial runs to Quorn with a rake of ex British Rail Mk1 carriages. The engine, No.5231, one of the LMSR Class 5s was featured in a big way during the first couple of years until it became in need of repair as a result of a superheater flue tube collapse. Owned privately No.5231's first claim to fame came on 30th September 1973 when it hauled the inaugural train between Loughborough and Quorn & Woodhouse.

These first services were operated by two engines and a couple of coaches as depicted. The reason for the somewhat extravagant use of two engines was because there were no run round facilities at the time at Quorn & Woodhouse. Therefore an engine had to be coupled to back and front of the train to enable it to run back and forth. Shortly after the start of services a temporary suspension

Other small locomotives arrived on the line and these included *Barrington* and *Lamport No.3* as well as two large, ex main line types *Boscastle* and *Duke of Gloucester*. The Great Central or Main Line Steam Trust as it was then known was fortunate in possessing a working main line steam locomotive which had been bought immediately after the end of British Rail steam working. It was transferred to

occurred owing to a problem with the Light Railway Order.

With the large stud of locomotives which the Great Central possessed and the fact that two of them were in need of repair, the next logical step for the project was to build an engine shed. A suitable site was chosen at the north end of Loughborough Central but the railway was careful to provide for

By 1976 regular steamings of locomotives and running of train services were taking place, attracting many visitors to Loughborough Central. Here,* King Haakon 7 *and ex LMS Class 5 4-6-0 No.5231 are seen being prepared in the yard for a day's work, alongside the newly restored signal box.
(M. Hooper-Immins)

any future decision to extend track northwards again to Ruddington should funds permit, as the artist's impression of the shed complex shows. At the same time as the shed was being constructed by a dedicated team of volunteers, British Rail decided to sever the link to the north of Loughborough, as far as the former Midland Railway, where a new junction was to be installed for East Leake trains. This meant that the Weekday Cross spur would disappear in Nottingham and also meant that the possibility of the GCR being rail connected was becoming more remote. Rolling stock deliveries would be confined to road transport which would have to unload at Quorn as no access was available at Loughborough.

By July 1974, after much difficulty, regular operation of steam trains began again on the Great Central between Loughborough and Quorn & Woodhouse and these were supervised by British

An artist's impression of the locomotive running shed and work shops at Loughborough.

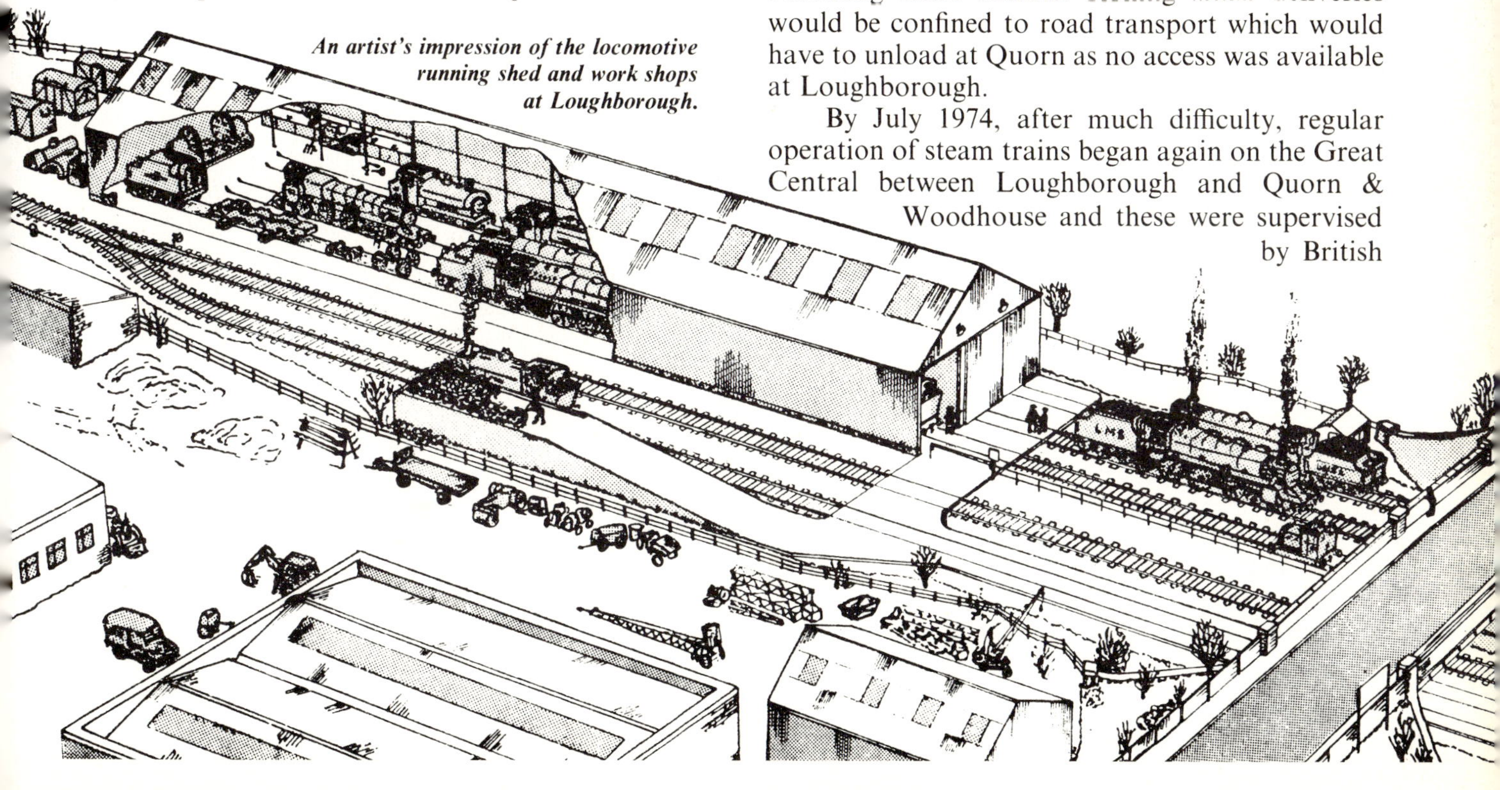

Railways' personnel. To keep the track in position with a view to its eventual purchase The Main Line Trust had to continue the monthly payments of interest, until sufficient funds could be found to purchase the track. Of the tasks the Main Line Steam Trust faced fund raising was probably the most daunting of all, as to secure the line between the two operating centres some £98,000 for one track alone was needed. To include the section to Rothley was going to require double that amount. The cost of purchasing as far as Birstall was out of reach at some £256,000, as were other options which would have made some two track operation possible. In connection with the fund raising appeal a small booklet was produced called the Main Line Steam Trust Development Appeal which contained a small amount of history of the line as well as why the Great Central deserved to be preserved and how one could go about supporting the scheme.

Another achievement in 1974 was the introduction of a timetable showing the services between Loughborough and Quorn & Woodhouse and as one can see it announces the extension of these to Belgrave & Birstall – an ambition which will very shortly come true.

The first alteration to this timetable was to come about as a result of the extension of services to Rothley which officially started in January 1976. This followed a trial week-end in September 1975 when Rothley had been used. The inaugural train left Loughborough Central at 12.15pm after a civic send off by the Mayor of Charnwood as well as the Mayors of Epinal (France) and Schwasbisch (West Germany) who visited Loughborough on twin town visits. The President of the Main Line Steam Trust, the Mayors and the other civic dignitaries who were present for the occasion dined with the railway's directors in the Gresley buffet car which was one of the Group's earliest acquisitions. The engine which pulled the train was *King Haakon 7* the Norwegian State Railways Mogul which was bedecked with the flags of the countrys of the visiting dignitaries.

GREAT CENTRAL RAILWAY

PASSENGER TIMETABLE

26th October, 1974 until further notice

SATURDAYS AND SUNDAYS

Miles			Morning				Afternoon						
			SX								Q	Q	
0	LOUGHBOROUGH CENTRAL	Dep.	11.30	..	12.30	..	2.00	3.00	..	4.00	5.00	6.00	..
$2\frac{1}{4}$	Quorn & Woodhouse	Arr.	11.40	...	12.40	...	2.10	3.10	...	4.10	5.10	6.10	...
5	Rothley	Arr.	..	..	..	..	..	..	..	..	..	..	..
$7\frac{3}{4}$	Belgrave & Birstall	Arr.	...	...	...	...	...	...	...	...	...	...	...

STEAM LOCOMOTIVES WILL HAUL ALL SERVICES AND THE TICKET INCLUDES ADMISSION TO LOUGHBOROUGH CENTRAL STATION WHERE A NUMBER OF RESTORATION PROJECTS ARE IN PROGRESS

Miles													
0	Belgrave & Birstall	Dep.	..	..	..	..	..	..	..	..	..	..	..
$2\frac{3}{4}$	Rothley	Dep.	SX	...	...	...	...	...	...	...	Q	Q	...
$5\frac{1}{2}$	Quorn & Woodhouse	Dep.	11.50	..	12.50	..	2.20	3.20	..	4.20	5.20	6.20	..
$7\frac{3}{4}$	LOUGHBOROUGH CENTRAL	Arr.	11.58	...	12.58	...	2.28	3.28	...	4.28	5.28	6.28	...

SX – Saturdays excepted. Q – not after dusk (lighting-up time).
Services between Quorn & Woodhouse and Belgrave & Birstall will be reintroduced in due course.

Additional trains may be run subject to demand. The Trust reserve the right to deviate from the above timetable subject to operational considerations. Trains do not operate after dusk on winter evenings. Refreshment facilities are available at Loughborough Central Station and a buffet car is usually included in the train formation.

Fares: Loughborough Central to Quorn & Woodhouse: 45p. return or 25p. single.
O.A.P.s & Children 25p. return or 15p. single.

MAIN LINE STEAM TRUST LIMITED

Registered No. 1024901 England. Registered Charity No. 263162

Registered Office: 13 NEW STREET, LEICESTER Telephone: LEICESTER 21443

The two days in 1975 not only made history by reopening the line to Rothley for passengers for the first time since 1963, but they also encouraged the MLST to press ahead with plans to secure the section to Rothley, thereby achieving another part of the eventual aim. Of course the extension to Rothley also encouraged more locomotives to come to Loughborough, and large ones at that, since the group now had a more realistic running length. The first of such locomotives to arrive was No. 6990 *Witherslack Hall*, which was purchased by the Witherslack Hall Society and moved to Loughborough from Woodham Brothers, Barry on 5th December 1975. The 'Modified Hall' class engines were frequent users of the Great Central main line when working cross-country trains over the Banbury to Woodford Halse link, but No.6990 has even more claim to fame than its sister engines as it took part in the Locomotive Exchanges of 1948, running along the London Extension. It competed with an LMS Class 5, a 'West Country' class from the Southern and an LNER B1 class, examples of which are all on show on the present day Great Central Railway. In this respect the major four pre-Nationalisation railway companies were represented on the former London Extension. No.6990 *Witherslack Hall* has been restored in GWR Brunswick green livery and was steamed in late 1986.

Around the same time two other locomotives arrived, including one from the Keighley & Worth Valley in the form of an N2 class 0–6–2T No.4744 which is owned by the Gresley Society. The second, owned by the Thompson B1 Locomotive Society being B1 No.1264 from Barry and which is still undergoing restoration at Loughborough. The N2 was luckier however in that after three years work it returned to service. Whilst on the Keighley & Worth Valley Railway the N2 shot to fame when it was used in the making of the film 'The Railway Children'.

To cope with the massive programme of restoration, appeals were launched by the various locomotive owning groups, but the real problem was whether the line the engines were to run on could be secured. In December 1975 British Railways announced that unless the line could be purchased outright by 1st April 1976 the track and ballast were to be lifted. It was soon clear that with the total cost, being some £279,000 the Main Line Steam Trust would not be able to realise this amount before the deadline. As a result the directors of the Trust decided to set up a brand new organisation, a public limited company, which was not surprisingly to be known as the Great Central Railway Company (1976) PLC. The purpose of this being to raise the amount needed by the sale of shares in the Company to the general public. British Railways at this news agreed to extend the deadline to 1st July 1976 and the new company share issue was launched in the May of that year.

At a deserted platform 1 Loughborough Central, **King Haakon 7** *waits to haul the short passenger train down the yard after running round into platform 2.* *(M. Hooper-Immins)*

The approaches to Loughborough Central in 1972 with the main lines in an advanced stage of deterioration. Note the traces of the old 'up' and 'down' loops, though the most prominent features are the bridge, booking office and water tower.
(M. Hooper-Immins)

The typical entrance to a Great Central island platform at Quorn & Woodhouse on 14th July 1974, having opened its gates to the public after a short break in services, pending the successful application of the Light Railway order. (M. Hooper-Immins)

Looking north to Loughborough Central from the station overbridge at Quorn & Woodhouse prior to the commencement of Main Line Steam Trust trains. In this picture the old 'down' main line is still in situ as well as the crossover. (M. Hooper-Immins)

Sadly, despite mammoth efforts to raise the money needed through wide media coverage the plan failed. The chances of seeing a stretch of double track being preserved were now at an end and it was only because members of the Main Line Steam Trust bought shares in the PLC that the company was able to purchase the line between Loughborough and Quorn & Woodhouse. After payment had been made for this section British Rail gave the Great Central a further six months to buy the section to Rothley, where the railway had started running to way back in January 1976. Although prepared to leave this section for preservation British Rail began removing the line and ballast between Rothley and Birstall as well as taking up the second track between Loughborough and Quorn, which was a bitter blow to the Great Central enterprise. To add insult to injury British Rail performed the work on the Loughborough to Quorn section first, during the July and August period which forced the railway to close down at its major operating season. Of the 725 yards of track which was recovered 165 was cut up for re-rolling by contractors whilst the rest was to be used elsewhere on the British Rail network. From the Rothley to Belgrave & Birstall section some 15,364 yards of track were retrieved and taken away to British Rail's London Midland Region depot at Lenton for re-use.

Turning back to the Great Central, 1976 was the most crucial period for the Company and to ensure that the section to Rothley could be retained an appeal was launched to encourage people to buy shares in the project. One of the Rothley station staff, in desperation, made a last minute notice with a tiny portion of rail attached to it to show how

Quorn & Woodhouse station in the early years of preservation seen from the 'up' side. On the right is the old yard which is somewhat overgrown and serving as a temporary car park. This area now sports a siding where the bushes are and acts as a showground and picnic site. *(M. Hooper-Immins)*

much one 25p share would buy at British Rail's asking price! During this appeal two locomotives featured prominently, namely Class 5 No.5231 and *King Haakon 7*, though the latter had an accident at Woodthorpe Bridge on one of its runs, during which a couple of the footplate staff were injured as

Quorn & Woodhouse station buildings and platform looking towards Rothley. The track at the end of the platform is almost hidden by weeds as the Main Line Steam Trust had not started restoration or running to Rothley at this time. *(M. Hooper-Immins)*

a result of a plug in its firebox failing. No.5231 became a regular star, attracting attention from both the major television channels as well as a number of newspapers and the locomotive even gained a name into the bargain. On one occasion two Chelsea Pensioners came to visit the line especially to see the 'Black 5' because it bore the name of their old regiment. In the same year No.5231 also carried the headboard of 'The South Yorkshireman' which was one of the named trains that used to run on the old Great Central until the withdrawal of titled trains in 1958.

In the months following the shutdown of the railway to allow track lifting to take place, the Great Central lost no time in setting about the task of providing a much needed run-round loop at Loughborough Central station. This followed the course of the trackbed through the 'down' side of the platform. With much effort and time put in by the volunteers the task was completed very quickly though in not the most favourable of weather conditions. One of the first trains to run over the new track was a special operated for the filming of a documentary on Thomas Cook by the BBC. The new loop in operational terms was a considerable asset to the Great Central as it did away with the need for costly and unnecessary shunting manoeuvres which in turn made tremendous savings. Indeed after the success of the new facility at Loughborough a decision was taken to provide Rothley with run-round facilities.

Rothley also saw the installation of another part of the past when the station was refitted with gas lighting after period fittings had been obtained from British Rail Collectors' Corner in London. These were assembled in some 18 places on the station and on 5th November 1976, they were officially turned on by Sir Mark Henig, Chairman of the English Tourist Board, while he was attending the Great Central's annual Guy Fawkes Party.

Gas lighting on Rothley station, February 1987.
(P. D. Nicholson)

Although the station at Rothley itself was beginning to revert to its original splendour the future of the track between the latter and Quorn was still in doubt. The share issue, which had been launched back in September to raise the £70,000 needed to complete the purchase of the track, was still short of the target despite an encouraging response. To boost interest in the scheme the Great Central was selling a limited edition of reproduction maps from Henry Fowler's survey of the course of the Great Central through Leicester. The first of these to be sold was presented by Lord Lanesborough to Leicester Museum, with the second copy being handed over to the National Railway Museum at York. By 26th November £163,000 had been raised towards the purchase of the complete line which left some £80,000 outstanding for the section to Rothley, but by Christmas the sum needed had been reduced to £64,000.

During the traumatic time of trying to raise the money to purchase the whole line, one historical item arrived at Rothley in the form of an old North London Railway carriage which had served as a greenhouse in a garden in Rugby until being rescued. This particular vehicle had three compartments, two of eight seats and one double of sixteen seats. Apart from this and the fact that inside the doors are written First and First Smoking, little is known about its history.

At the start of 1977, on 4th January the Great Central was given a stay of execution until 20th January by which time it had to raise the money and hand it over to British Rail, or else the track between Quorn and Rothley would be lifted. On the 4th the Great Central was just £50,000 short of the £2.25 million as a result of support given by well wishers of the project. At last, with the help of a bank loan and two interest free loans from Leicestershire businessmen, the cheque for the outstanding money was handed over to British Rail on January 19th, with just one day to spare before the deadline. In this light the future of the five and a half mile run to Rothley from Loughborough was now secure and it also meant that once the Great Central had been granted a Light Railway Order it would be possible for the volunteers to run the railway without the supervision of British Railways' staff as had been necessary since 1973.

The next special event to occur for the Great Central was a visit from His Royal Highness Prince Richard, Duke of Gloucester who visited the line on the 13th July as part of a select tour of sites in his capacity as President of the East Midlands Tourist Board. Preparations for this visit had been started much earlier with the marshalling of a special train consisting of three coaches, namely a first open, a Gresley buffet car which had come from the old 'Flying Scotsman' and a brake coach. The first open coach had been acquired for operations at the beginning of the Easter period to provide extra accommodation for meal sittings, and of course the coach came in very handy on the day of the Royal visit. This occasion marked the turning point for the Great Central catering which will be discussed later in this book. Aside from the coaching stock, the stations, locomotives and other items on the line were cleaned and polished for this one special occasion. The Duke was carried on the train to Rothley, hauled by *King Haakon 7* which departed from Loughborough at 1 o'clock which was to be the standard time for the departure of luncheon trains in the future. On arrival at Rothley the Duke was introduced to the President of the Railway, Lord Lanesborough, to the Chairman and to the Chairman of the 71000 Duke of Gloucester Steam Locomotive Trust. After this the train returned to Loughborough, though on the way a stop was made on the Swithland Viaduct where a five course meal was served, consisting of spring soup or avocado mousse; chicken kiev with garden peas, carrots, new potatoes and croquettes or 'Cold Collation ·reat Central'; peaches cooked in brandy or strawberries and cream; cheeses of the Midlands followed by coffee. Before the Duke left he formally agreed to rename the locomotive named after him, once restoration was finally complete. Originally the locomotive had been named after his late father when he was President of the International Railway Congress.

The next major event planned was to bring a large, three cylinder locomotive to Loughborough, which had run on the Great Central back in 1962. Two of the Railway's directors had bought what is regarded by many as probably the best preserved locomotive in the country, namely No.5690 *Leander*. Purchased for a sum in excess of £50,000, sadly the engine was not moved to the GCR due to various political problems being encountered.

Whilst the transfer of *Leander* might have been the solution to the railway's financial problems, in that the locomotive would have attracted more revenue, it was left up to the Charnwood Council to provide the solution to the ever mounting debts being incurred by the Great Central. This was achieved by purchasing the land from Loughborough to Birstall and obtaining a Light Railway Order with the land being leased to the Great Central for £1,750 per annum. At a special ceremony two pieces of paper – a lease for the land on which the track stood on and the Light Railway Order from the Department of the Environment – were handed over, thereby providing the Great Central with the necessary security to operate and develop a train service. To reach this historic and magic moment had taken some nine years hard work by the stalwart volunteers who wanted the railway to survive for posterity. On the occasion of the handing over, the shareholders and Charnwood Council officials heard the Mayor of Charnwood joke, that if the railway failed it would be taken over by the Council and the Mayor would become an engine driver while the Chief Executive would become a stoker!

During the 1978 season the Great Central was well on the way to becoming a major attraction in the East Midlands area after a rather inauspicious start. Some 2,500 people were estimated to have visited the line at this time. Visitors were able to sample rides behind a locomotive that had originally been deemed unfit to haul passengers because of its specifications. However the workshop at Loughborough had changed the situation and given it a new lease of life by rebuilding No.39, a Robert Stephenson and Hawthorn tank of 1938. The work involved retubing the boiler, replacing the saddle tank with side tanks and the installation of vacuum brakes for passenger working.

In addition to No.39 another engine joined the operating fleet and this was the N2 No.4744 which until mid 1978 had undergone a major restoration

The now unique Gresley N2 0–6–2T No.4744 (BR No.69523) at Loughborough where it is seen running round its train during the Easter 1986 services. ***(Author)***

programme. This brought the total of operational engines to four, the others being *Robert Nelson* and *King Haakon*. It also proved to be a good year for the Great Central's dining trains 'The Charnwood Forester' and the 'Carillon Luncheon Train' with patronage being up on the previous year. Another new feature which was attracting attention was the introduction of a schools week where the Midland Newspapers reported that some 3,700 children attended on the three days set aside in the May of that year.

June 1978 saw the arrival of yet another large locomotive which was brought from Carnforth by road. No.1306 *Mayflower* had for several years worked specials, either on her own or with other engines from Carnforth before being brought to the Great Central. It is a fine engine and the Great Central was lucky in getting her when they did as it was to make up for the lost opportunity with *Leander*. After her arrival it was hauled down from Quorn to Loughborough where work was started on a full scale overhaul before she was put into service. Before the year was out *Mayflower* was proved to be not the last of the new engines to be delivered that year. Thanks to the help of a Nottingham businessman and volunteers from the Great Central Railway another locomotive was added to the stud at Loughborough. Unlike *Mayflower* which had been in heavy use until its transfer to the Great Central No.5224, an ex Great Western Railway 2–8–0 tank engine was in somewhat poor state having been allowed to rot in the 'Graveyard of Steam' at Barry in South Glamorgan.

As well as progress in the Locomotive Department, evidence of the achievements in other directions on the railway were apparent too. The Signal and Telegraph Department which had been in existence for five years with the sole aim of re-establishing a proper and comprehensive signalling system on the railway, opened the first part of its gargantuan task, namely the signalbox at Loughborough which had been stripped down to just a brick and wooden shell when it was closed in 1965.

B1 class 4–6–0 No.1306* Mayflower *arriving at Quorn & Woodhouse on 18th July 1978. Purchased by a Mr Gerald Boden it had previously been based at Steamtown, Carnforth. (M. Tye)

GWR 2–8–0T No.5224 acquired by Mr Roger Hibbert from Woodham Bros, Barry, arriving at Quorn & Woodhouse on 25th October 1978. Transport was by Leicester Heavy Haulage whose yard was conveniently located alongside the GCR at Loughborough. (M. Tye)

One of the earliest tasks of restoration was the signal box which had been gutted beyond all belief after abandonment of the line by British Rail. It lost its lever frame and all the windows which had been restored by the time this view was taken. A new lever frame was brought from Ruddington and was installed ready for connection to the signalling apparatus, which was later laid and commissioned to control the station area around Loughborough. *(M. Hooper-Immins)*

Work on the signalbox included the installation of a 50 lever frame which was taken from the similar box at Ruddington. The original lever frame had been taken out and broken up for scrap. In order to fit the new frame various alterations had to be made as it was too large for the box. Other features included the reglazing of all the windows, new floors both up and downstairs were installed and a new chimney was also erected. For housing the electrical relays a large cabinet had been transferred from Ruddington along with the frame. Other electrical features in the box included a 50 line telephone exchange which is now used to control the telephones in the station area. During 1978 in connection with the telephone apparatus installation, work was carried out on re-establishing the telegraph wires between Loughborough and Rothley in order to give the Railway a better communication system. Most of the telegraph poles along the route had to be replaced and there was no wire left on the poles as this had been stripped by British Rail when the line was abandoned. During 1978 also, the Signal and Telegraph Department installed two ground frames, one at Rothley and the other at the south end of Loughborough Central. Both of these, were at the time, released using Annetts keys carried along with the staff on the train.

In general the Great Central made much progress during 1978 in establishing itself as a major tourist attraction within the area. This was an incentive to the volunteers working on the station and rolling stock restoration projects. At Loughborough much effort was put into the establishment of the new museum which had formerly been the information and appeals office. Also, work was centred on the Ladies/Waiting Room and the overall canopy. Simultaneously work at Rothley was being carried out to improve the appearance of the island platform with the main centre of attention being the stairway and the ticket office.

One of the main benefits for the Great Central was the lifting of the ban on operating after dusk which meant that the last train was put back to 4.05pm instead of 2.50pm. Special advantage of the late running was taken on 4th November of that year when a train left Loughborough at 7.0pm for Rothley in connection with a Guy Fawkes party. This train proved so popular that many passengers were in fact turned away. Three special night time trains were operated by the catering department in

October, November and December. In addition to these a charter special was operated for a printing company which allowed 60 diners the chance to sample a champagne reception, and a dinner including smoked salmon and roast duck served in the vicinity of the Swithland Reservoir. The diners were entertained at Rothley and Loughborough by the World Championship Ratby Brass Band.

During December 1978 the Great Central operated its usual Father Christmas Specials which pulled in record crowds of people to see Santa on the four days of operation. Also in attendance were his fairies, pixies and tweedle dee and tweedle dum who served mince pies and sherry. The Santa trains were operated in the main by No.4744 with the exception of one which was hauled by No.4 *Robert Nelson*.

One other development made on the Great Central was the reforming of the Carriage and Wagon Department to cope with the repairs and restoration needed on coaches to provide enough seating capacity for passengers. The first work to be carried out was the painting of a brake van in BR brown and the fitting of a vacuum pipe. Shortly after the latter task was completed there were three additions to the fleet: the second Gresley Society coach, a brake third; an LNER 6-wheel parcels van No.70654 and a 'Catfish' ballast hopper wagon. The 6-wheel van was overhauled and turned out in red and its dynamos were serviced. The 'Catfish' was repainted in black with white lettering. By the end of 1978 the Great Central had some six carriages in running order, being formed of four ex BR Mk1 carriages and a couple of ex BR Gresley buffet cars. To cope with the obvious increase in traffic though, a further four carriages were acquired from the Nene Valley, which arrived on the Railway in 1979. The four coaches arrived from Peterborough on 10th April and after a quick programme of repairs two of the latter were able to enter service in time for the Road/Rail steam event that year. Even though a rake of eight coaches was provided on both Easter Sunday and Monday, many passengers still had to stand.

The now annual Easter Road/Rail steam event has proved to be very popular, boasting thousands of visitors who come to travel on the Railway, view the various traction engines on show, the threshing machine and enjoy the fairground amusements, trade stands and locomotives which are on display. Just prior to the 1979 display a $7\frac{1}{4}$in gauge miniature railway was established around the former goods yard at Quorn having about 500ft of track and a 2-car battery electric unit, though sadly this scheme never really took off and the line was dismantled a few months afterwards. The station itself saw much enthusiastic labour expended on it with the roofing of the waiting room and the establishment of an electric supply to the main buildings. At around this time plans were put forward and subsequently implemented for turning the old goods yard into a car park and picnic area.

Work on the stations at Rothley and Lough-

Parcels van No.E70654 stands in Loughborough Central yard in February 1979 receiving attention in preparation for restoration to British Railways' Maroon livery with black ends and yellow and black lining. ***(M. Hooper-Immins)***

borough mainly centred on general repairs and the creation of a new Museum respectively. Throughout 1979 work was mainly carried out on the canopy at Rothley which had been a victim of dry rot and wet rot while the Museum was decked out with numerous exhibits in readiness for its re-opening. These mainly consisted of lamps, shedplates and bridge plates which had been lovingly restored, posters and various signs from stations of yesteryear. In addition to these were a collection of small items housed in a large display case plus a colour light signal and lever frame provided by the Signal and Telegraph Department.

On the traffic side four industrial locomotives left the Great Central line, these being an austerity 0–6–0ST which went to the Midland Railway Centre with one of the Sentinels, the other Sentinel moving to Steamport, Southport and *Lamport No.3* which moved to Shackerstone. As these engines were no longer part of the passenger operation their departure caused little upset to the success of the railway during 1979. The summer period showed a fairly good level of patronage with previous overcrowding of trains being eased by the introduction of the coaches from Peterborough.

Virtually all the trains during the latter part of the year were hauled by *Mayflower* which proved to be an extremely reliable locomotive after its introduction to service in the May of 1979. On Saturday 3rd November the Railway was taken over by the Army for the day and was operated by 275 Railway Squadron under the watchful eye of the Great Central staff. All aspects of the operation, including the driving of the engines were carried out by the soldiers. Two train sets were used, one hauled by No.1306, the other hauled by No.39. The two-coach rakes comprised of the FO, BSK and Gresley buffet pulled by No.39 and the second was made up of the TSO 3 × CK, SK and BZ pulled by the B1, No.1306. *Mayflower* was also responsible for a breathtaking performance on the Bonfire Night train when it ran at the head of a nine coach train. Throughout the rest of the winter period and into the spring of 1980 only a small rake of coaches was used in the hope of reducing fuel consumption in the interests of economy.

Former industrial steam locomotives destined for alternative homes. Seen at Quorn & Woodhouse on a winter's day in 1978 these are left to right: Hunslet 3193 of 1944, minus saddle tank and cab, and Sentinels 9373 and 9370, both built in 1947.
(M. Tye)

The major events of 1980 were the Road/Rail Steam at Quorn which gained recognition by the National Traction Engine Club as being the first official rally of the year. As usual a high turnout was witnessed on both days with the usual machinery on display, though the provision of GCR 'Director' class 4–4–0 *Butler Henderson* coupled to a 'Barnum' saloon acted as a crowd puller. Prior to the Easter Weekend another attraction was inaugurated with the opening of the Museum at Loughborough Central, boasting a fine display of exhibits which have been added to over the years. The new museum was a major benefit to one of the Railway's other annual institutions namely that of Schools Week which is held during the second week of May. The Museum provides the children with the chance to learn about the history of the Railway old and new, and the purpose of various exhibits around the walls. During 1980 some 5,216 children visited the Railway in one week and travelled to Rothley behind *Mayflower* and No.4744. Other locomotives were on display in the form of No.377 *King Haakon 7*, *Duke of Gloucester* and No.39. Schools Week is renowned for bringing out a new atmosphere on the Railway as regular workers become guides for anything from half a day to four days before returning to the Permanent Way or Signal and Telegraph Departments.

Mayflower ***heads a schools special, May 1979, this being the first occasion the locomotive saw service on the new GCR. (M. Tye)***

The next event in the calendar was the Railway's annual Bonfire and Fireworks Display held at Rothley and organised by local members of staff. Such was the demand for this spectacle that a special train was laid on at Loughborough to convey people to Rothley. Following the bonfire night bonanza, the Railway turned its attentions to the annual visits of Santa Claus by running Santa Specials on the 13th/14th and 20th/21st December. On the trains which were decked out with tinsel and other baubles, gifts were handed out to children while sherry and mince pies were served to the adults, despite some concern from the Catering Department! On the whole 1980 proved to be a good year for the Catering Department with an increase in the number of 'Charnwood Forester' runs as well as many special charter runs. In addition there were four special 'theme' trains; the 'Fifties Flyer' complete with Rock and Roll disco, the 'Forties Express' evoking wartime travel and the 'Thirties Limited' on which train a revue was performed and the 'Edwardian', on which a six course Christmas banquet was served.

1981 saw few changes in operation apart from the continual enhancement of the stations at Loughborough, Quorn and Rothley. Loughborough saw considerable work on the booking hall area with general repairs and revarnishing of the panelling. At the same time, work was being carried out to replace broken glass and to re-putty and paint the canopy above platform 1. Canopy work was also being continued at Rothley where new barge boards were fitted.

On the locomotive side the Great Central was lucky enough to be able to negotiate the loan of an engine from the National Railway Museum through the courtesy of the keeper, Dr John Coiley and his assistant John Bellwood. The engine concerned was ex Great Northern Railway Stirling Single No.1 which was for a short period, prior to making a trip to Germany in late 1982. With the arrival of the Stirling Single a special weekend was organised on

Right, top: Occasionally the Great Central plays host to visiting engines from other railways or museums and here Great Northern Railway, Stirling Single No.1 on loan from the National Railway Museum is seen departing from Loughborough Central with a train for Rothley on the 2nd December 1981. (M. Hooper-Immins)

Right: Another view of the Stirling Single as it waits to depart from Loughborough Central. In the background are two of the Great Central's ex LNER Gresley buffet cars NosE9122 and E9124. (M. Hooper-Immins)

G N R
Nº1

the 5th and 6th of December. The engine operated a normal service but because it was only able to haul a limited number of coaches, seats were available by pre booking only.

Another engine from the National Railway Museum, but on a long term loan is No.506 *Butler Henderson* which, after many months of restoration

The end of the line! No.506 **Butler Henderson** *and N2 No.4744 have just pulled into the platform at Rothley with a Road Rail Steam Special. Already by Easter 1986 work was progressing on the extension southwards to Belgrave & Birstall for which many track panels and other equipment has been amassed in the yard.* *(Author)*

work, was finally completed and saw service at the beginning of Spring 1982. To mark the occasion a special London & North Eastern Railway weekend was put on with B1 No.1306, N2 No.4744, GNR No.1 and GCR No.506, all in steam. On this occasion a different locomotive was provided in steam on each train with the first and last trains being double headed. Such was the interest aroused by this weekend that it was decided to repeat the event for future years, although traction for these days has tended to be of the diesel nature.

On the 8th/9th May, with GNR No.1 operating the trains, a Rail Letter Service was introduced with the travelling Post Office equipment acquired from Railway Vehicle Preservations Ltd. The operation of the lineside equipment at Quorn & Woodhouse by No.506 *Butler Henderson*, provided a remarkable sight and was witnessed by the Head Postmaster of Leicester and Loughborough. To commemorate the event a special first day cover was designed and a hand stamp approved by the Post Office for all mail posted that day. Also issued on that day was a complete series of Rail Letter stamps depicting GNR No.1 and LNER No.4744 in full colour.

The main aim of the Great Central 1976, like the original company bearing that name, has always been to provide value for money and cater for all types of people. In a bid to encourage and allow handicapped persons to enjoy the facilities, a lengthy fund raising and conversion/restoration programme was embarked on to provide a coach suitable for disabled passengers. The project was completed in time for the launch on 2nd April 1983 which was attended by a group of VIP disabled visitors.

Around the same time the Carriage and Wagon Department was fortunate in acquiring covered accommodation from Derby, in the form of a shed kindly donated by the British Gas Corporation. This was subsequently dismantled and moved to Loughborough and Rothley for storage.

One new feature during 1982 was the introduction of the 'Leicester Mercury' 'Funday' on the Great Central Railway which was attended by some 2,800 people, providing an enjoyable day out for all concerned. Since its inception this event has proved very popular and is repeated every year.

1982 proved to be a boom year with an increase in loadings by no less than 21.89% on the previous year with 128,470 passengers travelling on the line in 1982, compared with 105,398 in 1981. This upward trend in patronage continued into 1983 with a 40% increase in revenue although the Christmas 1982/3 period provided a slight trough. Figures gradually picked up for the Easter period which was mainly due to the extra activities organised around this event.

Among activities of note were the television filming by the BBC for C. P. Snow's 'Strangers and Brothers' and the showing of Central TV's film of the Great Central Railway on 7th February. Another notable event was performed to celebrate the beginnings of Thomas Cook's Travel Agency in Leicester. The present firm bearing that name decided to utilise the Great Central Railway and the East Midlands Airport as the venue for entertaining in excess of 100 travel agents.

Aside from the usual events, plans were being put into force during 1983 to create a pathway alongside the track on the course occupied by the former 'down' main line. However, much depended on the future plans of the Railway such as the relaying of the second track, the provision of a run round loop at Quorn and the Rothley to Birstall extension. One further scheme was considered along with the Railway path, namely that of extending the embankment north of Loughborough with a view to connecting up with the main line again. Most of the emphasis was put on the restoration of the Quorn Loop in view of the large number of requests for the project to be instigated which it was estimated would cost in the region of £25,000. For the purpose of raising finance a fund book was opened for the purchase of one inch of track at £1.50 a piece.

In addition to the various new schemes planned at the time for enhancing the facilities of the Great Central, the Company reached the tenth anniversary of running its first train. This had been run officially on 24th June 1973. The occasion was marked by a special train from Loughborough to Quorn where a clambake was held. To commemorate this important milestone in the Railway's history a barbecue was organised at Quorn for 24th June 1983. On the following Saturday and Sunday the annual LNER gala weekend was held which was the last special event prior to the autumn season. This year saw a new venture established on 24th/25th September at Quorn in the guise of a Crane Weekend including the Railway's breakdown crane RS 1097/45, two crane traction engines and some modern high capacity diesel engined cranes.

By the end of 1983/4 several significant changes occurred on the Great Central with the resignation of the Company's Chairman and General Manager. To fill the gaps five of the directors, L. T. Blower, G. Boden, K. J. Kellett, R. J. B. Lovatt and B. H. Porter were given the task of administering the day to day operation, with Mr. G. Boden being appointed Managing Director to co-ordinate schemes. The principal task of these directors was to ensure that the Railway operated safely and efficiently and was commercially viable. As far as the financial side was concerned the most important concern was the removal of the debt to the bank. In addition to the latter consideration efforts had to be made to promote the Railway and its attractions to the public at large. The Catering Department was reorganised and as a result saw a lot of new business generated. To complete the picture the locomotive restoration projects were getting nearer to being steamed, ready for the expansion of operations with the Quorn Loop.

By Easter 1984 preparatory work for the installation of the loop commenced which is intended to allow additional trains to run at busy periods as well as to allow the demonstration of mail and freight trains without disturbing the normal service. The need for the loop was made apparent by the popularity of the Road/Rail Steam event which, due to the number of visitors on previous years, was extended to operate for the first time on the Good Friday. The following month another first happened for the Great Central with the introduction of a new timetable allowing the running of two train sets on a regular basis. This allowed the service frequency to be improved with one train every 50 minutes instead of every 70 minutes. This was made possible mainly as a result of the commissioning of stages

Belgrave & Birstall station after closure of the line as a trunk route in 1966. The section between Rugby Central and Nottingham remained open and the station shuddered with the passage of the occasional DMU from Leicester on its way to Loughborough Central. In this scene the buildings are in fairly good condition though all the fittings, windows and doors have been removed save for the station nameboard under the canopy. Note the extra building on the platform, the last but one from the end.
(M. Hooper-Immins)

one and two of the signalling scheme. The new timetable meant that as soon as one train had arrived at Loughborough from Rothley another was ready to depart for there. From the point of view of interest the running of two train sets means that there is always one rake of coaches at Loughborough at any given time.

Throughout the rest of 1984 much progress continued to be made in reducing the overall expenditure and cutting down on operating costs to help ease the burden of the bank loan. In order to ensure that the Railway ran to a closer budget than previously, only essential expenditure was permissible which meant the shelving of certain projects. For example the Carriage and Wagon Department's plan to bring the Wood Green 'Coronation' shed to the Railway, has had to be postponed.

Fortunately the cut-backs did not affect the development of the Quorn Loop project which enjoyed a huge response to the fund raising appeal. Whilst the necessary sums were being amassed, much work was carried out on completing the trackbed preparation and revamping the drainage system which had been defunct for many years.

Although expenditure on a large scale was restricted, the Railway Board did not overlook its future plans and intentions, by pushing ahead with the extension of the Railway from Rothley to Birstall. The main reason for wanting the extension was to allow the Great Central to get the best out of the Light Railway Order which covered operations between Loughborough and Birstall. The completion of the extension (which is underway at the time of writing) will add an extra $2\frac{1}{2}$ miles of track, bringing the total running line to $7\frac{1}{2}$ miles and reaching the outskirts of Leicester. During Autumn 1984, once the busy season had slackened off a survey was carried out of all the structures along the new route, and these were found to be in a satisfactory state. At the same time estimates were prepared to ascertain the cost involved for the whole project. Fund raising to cover the expenditure did not begin until some time later although clearance of the embankments and the trackbed of the many trees which had sprung up following closure, was commenced soon after the inception of the Birstall Project. Along with the decision to reach Birstall the Great Central Board of Directors was negotiating with local authorities with a view to extending the railway northwards to Ruddington.

As 1984 neared its end the Great Central had

enjoyed another busy year with the usual events at Easter, Schools Week, Special Weekends and the Bonfire Night and Santa Specials all proving even more popular than ever. The turn of the year saw a major impetus as far as the Quorn Loop was concerned. On the weekend of 19th/20th January 1985 more than 30 members installed a point to the north of Quorn. The work, which began at 4 o'clock in the morning and caused the suspension of services that day, involved the digging out of the main line ballast and lifting of the track panels, ready for the turn-out which was laid and ballasted. The track taken out from the main line was then laid to form the first part of the actual run-round. To control the turn-out a ground frame was installed next to the main line, though when the signal box at Quorn is finally completed and placed on a brick base, the frame will be dispensed with. To help with the work No.25 *Northiam*, an Austerity 0–6–0ST on loan from the Kent & East Sussex Railway was lit up. No.25 was a regular performer on the 1984 specials and various other services until it returned to its home on 1st May. Over Easter, the previous week saw the laying of the track into the 'down' line platform at Quorn in a dedicated effort to provide the chance for an exhibition coach to be placed in the station. This coach, the GCR 'Barnum' was used for showing the Railway's promotional video and housing the Museum's display.

Whilst the Quorn Loop thrived other projects saw little development or made no headway at all. On the subject of the Ruddington Extension, Rushcliffe Borough Council approached British Rail to discuss the possible purchase of the former GCR line from Ruddington to East Leake and negotiations are still continuing in 1987. The plan to build a footpath alongside the Great Central main line however, was abandoned by Charnwood Borough Council due to the need to reduce expenditure on MSC schemes. Even though this scheme, between Loughborough and Rothley, has been shelved, once the Birstall Extension is completed Leicester County Council intends to establish a similar facility between Rothley and Birstall, emulating the paths already set up on the old Great Central to the south of Leicester.

In order to co-ordinate efforts on the Ruddington and Birstall Extensions the Railway appointed a projects manager, as part of a general rearrangement of the management into a committee responsible for the acceptance of new locomotives and rolling stock onto the GCR.

The 8th July 1985 was a red letter day as far as the Great Central's Birstall Extension was concerned as the Duke of Gloucester came to inaugurate the Birstall appeal and the extension to the outskirts of Leicester. The occasion commenced with the Duke meeting a reception committee comprising of some of the Railway's important people after which, the Duke was led over to platform 2 to view the progress on the then virtually complete No.71000 *Duke of Gloucester*. The Duke himself showed a considerable interest in the locomotive and made a close inspection of the smokebox, entered the cab, and even climbed on to the tender. Following the inspection of No.71000 the party crossed over to platform 1 to board the two coach royal train hauled by No.506 *Butler Henderson* carrying the headboard 'The Carillon'.

The train made a slow trip along to Rothley where a number of guests were assembled ready for the opening ceremony. The train progressed south of the station at Rothley to allow the Duke to disembark from the train by means of a set of specially positioned steps. Following a speech made by the Railway's Chairman a chromium plated bolt on a dish was handed to the Duke who also made a short address. The bolt was then ceremonially placed in the fishplate which connected the existing track to the newly laid panels for the extension. This connected a further 80 yards of track in all to the already established main line, having been laid and ballasted during the weekends before the event. After the track was bolted together the guests and dignitaries watched the Duke climb to the footplate and drive No.506 through a banner declaring 'Steaming to Birstall', before retiring to a marquee in the station compound for a buffet lunch.

The extension was now started and all that remained to do was to amass the necessary funds amounting to some £246,000 to ensure completion. The appeal was very successful and in the first few days £9,000 pounds was amassed along with a guarantee of more than £1,200 per month coming in from the share scheme. The fund received a massive boost with an allocation of no less than £88,000 towards materials for the project.

At Rothley the present terminus of operations, plans for a shed are under consideration which would result in the need for development of a new car park south of the permanent way buildings. The provision of a depot/shed will allow for a Birstall–Rothley shuttle to be run once the $2\frac{1}{2}$ mile extension is complete. Expansion is now the key word and moving further up the line to Swithland it is the Company's intention to build a halt on the site where the original Great Central company had intended to erect a station. The reason for development of the Swithland area is because of its surrounding beauty, and now that there are no longer sidings connected to the granite quarries at Mountsorrel, the yard would make an ideal spot for a picnic site, amusement area and refreshment facilities. Eventually the Road/Rail Steam and annual bonfire will be transferred to this site as Quorn & Woodhouse is to be used for the erection of the carriage shed, which is currently in store. This will be extended, with more track being provided in the yard to prevent shunting movements from disrupting main line services and other associated facilities provided. In addition to the Derby shed, the 'Coronation' shed is still being considered as a possible aquisition, depending on whether sufficient funds are available for dismantling and transportation.

Aside from the sheds several other projects are now in hand. First of all the operation of a multiple train service is fast becoming a reality with the completion of the Quorn Loop. Now that the track is laid and connected up, all that remains to be done is the installation of the signalling and signal box equipment. The second interesting project is the construction of a landing stage north of the bridge beyond the present engine shed, which will provide access for disabled visitors to the GCR via the Grand Union Canal.

On the locomotive front 1986 was a bumper year with the introduction of three new locomotives to the already large stud of steamable engines. The first of these was No.68009 which was reintroduced on 16th March 1986, No.6990 *Witherslack Hall* was due to enter service at Easter for the Quorn Road/Rail Steam but did not appear until June. No.71000 *Duke of Gloucester* made her first run on the 25th May, much to the delight of the railway enthusiast fraternity, though as the engine was missing one set of coupling rods it was running with a 4–4–2–2 combination, as opposed to a 4–6–2 wheel arrangement. The coupling rods were finally delivered and assembled and the Duke was ready for service save for painting and lining out. As the locomotive was required for a special trip on 16th July for Kodak and Rushcliffe Borough Council, for whom a 'Charnwood Forester' train had been organised, it was given a preliminary coat of paint. A number of trial runs were undertaken the day before the event, though during the first of these the engine set fire to the bank just south of Loughborough Central causing a minor disruption to the afternoon's proceedings! Fortunately all went well the following night.

In order to ensure the smooth operation of the railway, various departments are needed, such as Locomotive, Carriage and Wagon, Catering, Permanent Way and Signal and Telegraph. All these sections, which are described in the next chapter have played a major part in establishing what is one of Britain's leading preserved railways.

Austerity 0–6–0ST, Hunslet 3825 of 1954 draws out of Quorn & Woodhouse with the 14.30 Loughborough – Rothley train, 28th February 1987. *(P. D. Nicholson)*

Chapter 6

The G. C. Today

On Train and On Shore Catering

Initially catering was not included as part of the railway preservation movement. This was possibly because people thought that this area would not necessarily be a suitable revenue earner, judged by the way in which British Rail then fared with its food sales! Until the early 1970s BR had made heavy use of catering vehicles, even on cross-country routes. A number of these were to become 'life expired', resulting in a plethora of rolling stock for acquisition by private railway companies such as the Main Line Steam Trust/Great Central Railway.

The first evidence of catering on the Great Central though happened more or less by chance because of the success of the opening of the station at Loughborough in the early days of restoration. To provide sustenance for visitors a temporary buffet was opened on the station for refreshments and light snacks. The latter proved to be very successful and so work was speeded up on the creating of a proper refreshment room, along with a shop which were both sited in the main station buildings. The refreshment room was ready for use at the start of summer 1973, coinciding with another significant event in the development of catering on the line. This event was basically to mark the inauguration of steam running to Quorn & Woodhouse, then the Railway's furthest limit. A special train was hauled by the Main Line Steam Trust's *King Haakon 7* consisting of a small Norwegian state coach which

***Inside the buffet at Rothley which was originally a waiting room but now serves a small range of light refreshments. The door at the back leads through to the waiting room past the ladies.* (D. Bonas)**

***A busy moment in the popular refreshment room at Loughborough Central which serves as a welcome resting place for railway enthusiasts and tourists alike.* (D. Bonas)**

took guests to Quorn & Woodhouse for a lobster bake. Much preparation had been done beforehand at the station, clearing long grass and assembling marquees. The lobsters were rushed by aeroplane across the Atlantic from Maine and a cook was supplied specially for the occasion. Disaster nearly struck as the refrigerated van bringing the dinner up the motorway broke down, but all was well in the end as the shellfish were taken to Loughborough and hastily put on a train to Quorn & Woodhouse. The lobster bake met with favourable comments and suggestions were made that the Great Central catering operation should be extended, but this could not be done until a suitable vehicle had been purchased.

A year later a number of Trust members got together and succeeded in acquiring Gresley buffet car No. E9124E which had been in service on East Coast Main Line expresses until it was withdrawn by British Rail in 1973. The coach was fitted out at one end with a buffet and had originally entered service in 1937 from York. Fortunately the coach had not been allowed to decay and after some minor attention it was put into service selling drinks. Similar catering coaches usually ended up being purchased by ambitious rail enthusiasts wanting to preserve something unusual. However with the area set aside for kitchens and buffets, seating capacity is limited so to make this type of coach a viable proposition the catering facilities have to be used. For the Great Central just running a coach with the odd drink being sold on board was not sufficient so plans were made to introduce a series of light, but hot snacks. Toast and baked beans became standard fare. In addition to the small snacks, the shape of things to come was demonstrated by the staff on certain occasions during 1975, when lavish lunches of an outstanding quality were served up to Civic Dignitaries touring the Railway.

It was not until 1976 that a real step forward was taken in the provision of on-train catering though, when more substantial meals began to appear, such as cottage pie and roast beef with all the trimmings. Most of the inspiration for this project came from the then Sales Development Manager of the Railway, Peter Ostle and two of his associates. Of course the Main Line Steam Trust's volunteers helped enormously in staffing the trains as to run a comprehensive catering scheme on trains requires dedication and hard work. The professionalism with which the organisation was being run was perhaps surprising as none of the volunteers had any previous catering experience, except for one member who was a catering student at the Loughborough Technical College. As time went by further improvements were made to the bar with the introduction of real ale in the form of Marston's Pedigree which was being served by gravity from the cask. Careful precision was called for as draught ale and trains in motion are not the best combination to have! Having catered for the real ale buffs and probably attracted more customers to the railway, a full bar and refreshment service could now be justified on every train. Mid-day trains now had a full restaurant service with three course meals provided. The latter was a first for the Great Central as it was the beginning of a boom in on-train meals. Other private companies did provide meals but they were not necessarily on moving trains.

For the Great Central and its catering staff 1977 was to be a busy year. The service was expanded with more dining accommodation being provided by a further coach purchased from British Rail, the broadening of the menu and the more frequent running of restaurant meal trains. The new coach was in superb condition when it reached the Great Central on 21st March 1977 having been obtained for the sum of £2,000. It was considered to be a valuable addition to the fleet, both because it offered a high standard of comfort and most important of all it could be used as extra dining accommodation during the forthcoming hectic months of celebrations brought about by the Queen's Silver Jubilee. The 'Silver Jubilee' dining train, to commemorate 25 years of Her Majesty Queen Elizabeth II being on the throne, marked another turning point for the Great Central catering service. This was the introduction of evening trains, the 'Silver Jubilee' being timed to leave Loughborough Central at 7.00pm. The menu on this service was quite phenomenal with six courses as shown below.

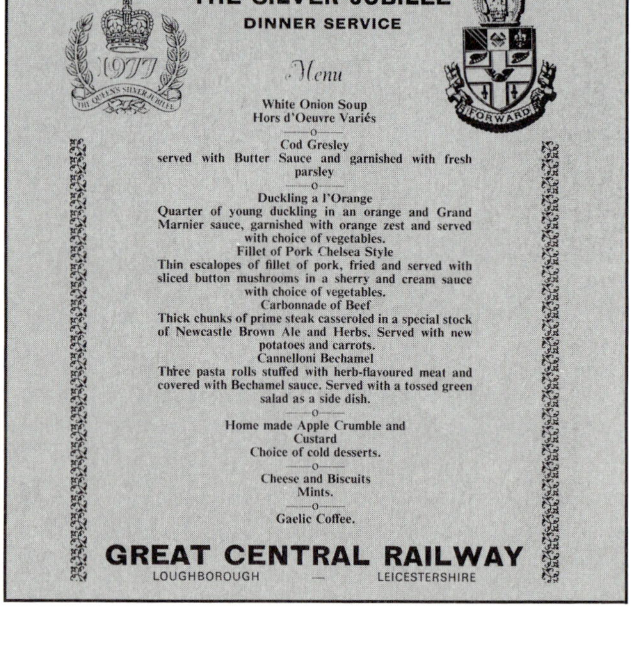
THE SILVER JUBILEE
DINNER SERVICE

1977
THE QUEEN'S SILVER JUBILEE

FORWARD

Menu

White Onion Soup
Hors d'Oeuvre Variés

—o—

Cod Gresley
served with Butter Sauce and garnished with fresh parsley

—o—

Duckling a l'Orange
Quarter of young duckling in an orange and Grand Marnier sauce, garnished with orange zest and served with choice of vegetables.

Fillet of Pork Chelsea Style
Thin escalopes of fillet of pork, fried and served with sliced button mushrooms in a sherry and cream sauce with choice of vegetables.

Carbonnade of Beef
Thick chunks of prime steak casseroled in a special stock of Newcastle Brown Ale and Herbs. Served with new potatoes and carrots.

Cannelloni Bechamel
Three pasta rolls stuffed with herb-flavoured meat and covered with Bechamel sauce. Served with a tossed green salad as a side dish.

—o—

Home made Apple Crumble and Custard
Choice of cold desserts.

—o—

Cheese and Biscuits
Mints.

—o—

Gaelic Coffee.

GREAT CENTRAL RAILWAY
LOUGHBOROUGH — LEICESTERSHIRE

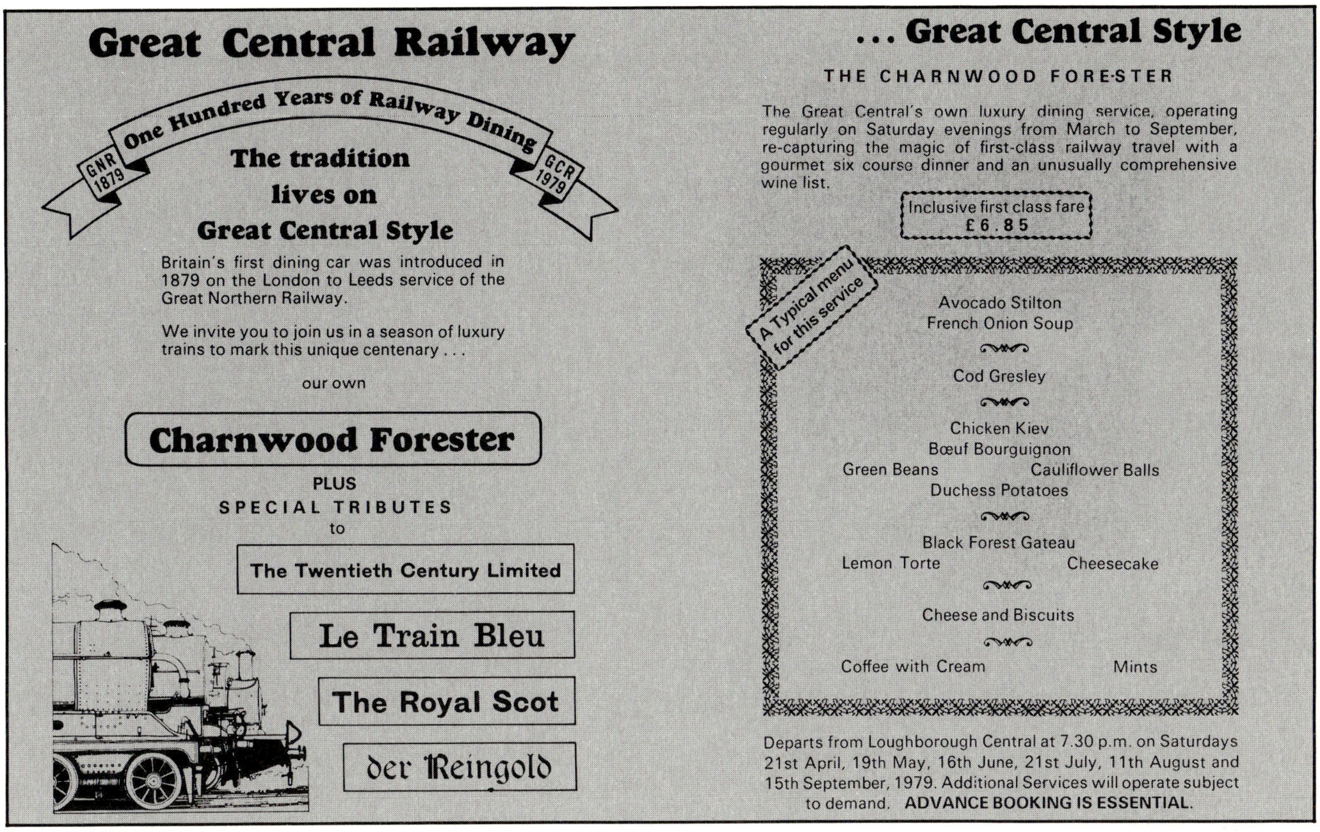

The Jubilee menus were an untold success and this prompted the catering staff to embark on creating a series of mouth watering restaurant menus for their evening trains. These were based on famous trains of the past such as 'The Rheingold', 'The Twentieth Century Limited', 'Le Train Bleu' and the 'Royal Scot'. In addition to these the Company ran its own special train called 'The Charnwood

16.35 on a Wednesday afternoon in Summer 1986 with the dining train standing in platform 2, being prepared for an evening 'Charnwood Forester' working. On the Great Central on-train catering is a speciality. *(Author)*

Inside the restaurant car with one of the tables laid out for the 'Charnwood Forester' train. Only the meal and the diners are missing! (D. Bonas)

Forester'. The latter was operated from March to September departing from Loughborough at 7.30pm on Saturdays. The first four trains mentioned continued to run for two years until the celebrations for the centenary of dining trains was over.

Besides the scheduled trains charter specials began to take off in a big way with trains being popular for birthday parties, office outings, weddings, dinner parties and other occasions. One customer, who took a big party for lunch and tea on the train actually wrote to the catering team saying that he thought they didn't charge enough, and sent £25 extra to buy some shares in the Railway! From my own experience meals on the Great Central are always very hot and served in a style befitting a top class restaurant.

In 1978 the Railway's catering services came to the attention of the East Midlands Area Tourist Board through the success that had been experienced from the programme of special trains the previous year. This recognition led the Railway to improve on its service yet again, even to the extent of having to cook breakfasts on the footplate! The latter stunt was mainly performed for a book on unusual meals sponsored by the English Tourist Board. As one leg puller put it, the extra cooking facilities on the footplate were probably needed due to lack of space on the stove in the kitchen car as a result of the heavy demand for meals.

In just three years the Great Central catering operation had come from supplying snacks and drinks to offering a full scale restaurant service with all the trimmings and a good selection of wines. Of course to cope with this the staffing situation had had to change dramatically to meet with the unexpected demand for meals on evening and weekend workings. Often, the volunteers who prepare meals on preserved railways are women but on the Great Central it has been dominated by males from the start. The first organiser, was an economics lecturer, Peter Ostle from Burton-on-Trent Technical College and he was helped by four volunteers. At the height of catering services in 1978 and the previous year, the Government's Job Creation Scheme provided the Railway with a full-time catering supervisor. On top of this the volunteer chef, Stephen Devine also became a full time member of staff after completing a catering course. In charge of the actual catering crew then was Marion Orton, whose husband had been the official British Railways' supervisor of the line when it operated under license from the former company. Every Saturday and Sunday throughout the year and on Wednesdays and Thursdays in July and August, work started early in the morning under Mrs Orton's supervision, laying the tables and preparing the food for later in the day.

At about this time a decision was already being made concerning the acquisition of a further coach in which meals could be prepared and eaten. The vehicle, No.1852 had been part of a pair of buffet lounge cars built for use in the 'Flying Scotsman' train sets. Incorporated in the design was a prestige feature of pressure ventilation and heating. In fact these were reckoned to be the first air-conditioned coaches built, prior to their introduction by British Rail 40 years later. Another feature of this type of coach was the inclusion of electric cooking facilities but these were later changed back to propane gas because the recharging of the banks of batteries needed for electric operation posed a problem. Much work had to be done on the coach before it could enter service. The exterior has been restored to almost its original state with teak panelling and a white roof, but because of the question of cost involved, several minor variations in the livery are evident when compared with the original colouring. However it is very striking in its teak livery when amidst a rake of 'blood and custard' coaches.

The seating area was changed as originally it only catered for 20 people. There was a large circulating area between the bar and the tables and at the end of the coach farthest from the bar compartment there was a separate section for ladies. The propane gas cooking facilities were retained by the Great Central although the rather sickly decor of pink and yellow formica panelling, which had been installed by BR when the coach was modernised during 1959, was replaced by wood grain panelling, more in keeping with the 1930's style. Both interior and

Restored No.1852 buffet restaurant car built at Doncaster in 1938 for the 'Flying Scotsman' standing out of service with other carriages awaiting repair or restoration. (Author)

exterior restoration was accomplished by the Loughborough firm of coach builders Willowbrook International, but much of the complex work on the vehicle's lighting, heating, plumbing and cooking equipment was effected by the workforce of the Great Central. After two years' work the coach was finally ready to enter service again in February 1980. The coach proved to be the centre of attention as far as quality of preservation went, but it had a far more important role to play than to be an item of rolling stock for admiration only. Many of its features were thought to be a tremendous improvement on the old buffet car which was long overdue for repair. These included better seating and lighting, the availability of mains electricity, the provision of a larger freezer and finally, more space for the preparation of vegetables. All this work including structural repairs had cost the Railway £12,500, though the coach was to prove a useful asset from the day it entered service in March 1980. Usually RB No.1852 worked along with the newly painted First Open No.M3042 which was turned out in blood and custard livery. The two coaches,

First open restaurant car No.M3042 complete with 'Charnwood Forester' 'Dining Car Express' carriage boards. The tables have been laid out ready for the 19.30 departure from Loughborough Central. (Author)

with a staff of chef, three waiters and a washer-up could cater for 80 diners.

As soon as the new Gresley vehicle had entered traffic the original buffet car No.9124 was withdrawn for repairs to enable it to continue in service as a back-up vehicle for No.1852, or for use in a second train. Eventual withdrawal from service for No.9124 came in 1981, which was to be a problem year for the Catering Department. The latter vehicle was used for the last time in July of that year when it became apparent that wet and dry rot were rife and nothing short of a complete rebuild could see the vehicle in service again. At around the same time, buffet car No.1852 began to fall short of expectations as several faults occurred. The first was the breakdown of the generator which required an overnight repair in order to keep the vehicle in service. The Great Central, because of this, was in dispute with the coach builders who had restored the carriage and this meant that it was not possible to effect the repairs that were needed. As a result alternative facilities had to be found because of the unreliability of No.1852 in the form of a new buffet car, which had the added drawback of requiring more finance, that was not readily available.

The next major advance in the history of the Catering Department took place in 1983 when a new First Open coach was delivered to ease the burden on No.3042 which was fitted out with a new carpet, as soon as the recent arrival, No.3126 could enter service. Before this could happen, damage to the carriage caused by a serious shunt had to be repaired which took nearly a month. Several other jobs had to be done including re-fixing all the seat frames, securing the toilet fittings and preventing several steam leaks. In addition the table tops were recovered and the interior was subjected to a massive clean. As the coach was to be used for dining services it was fitted out with an electric circuit for table lamps.

By autumn of 1983 the problem of finding a new buffet car was solved with the purchase of No.E1525 which was brought to the Railway in September. It entered traffic shortly after arrival having had a full overhaul to its catering and electrical equipment. The arrival of this coach meant that No.1852 could be taken out of traffic altogether for major interior and exterior repair work, ending the headache of having to carry out minor repairs in traffic. At the end of the year another vehicle bearing the number 1852 was acquired for the catering department in the form of a Restaurant Miniature Buffet, for use on a second train for the sale of light refreshments. Following a series of minor repairs and the installation of a new boiler the carriage entered service in Easter 1984 at about the time when the purchase was going ahead of a diesel multiple unit and trailer buffet car.

The evening 'Charnwood Forester' trains had now built up quite a reputation and were usually fully booked months in advance, due to the quality of fare provided by the catering staff under the newly appointed catering manageress Mrs Pugh. With the success of the evening named trains it was decided that the lunch-time meal train should be given a title. After a while a series of nameboards were prepared for the locomotive and rolling stock working the 1 o'clock departure to Rothley, proclaiming the name 'The Carillon'. This was suggested by Charnwood Council who maintain a bell tower in Loughborough called the Carillon.

From 1984 to the present day few changes have occurred to the Catering Department save for there being a change in managership, the department now being under the Commercial Manager, Terry Stirling, while the cooking is provided by Mary Brownlow. The other development is the arrival of No.E1526 sister coach to No.E1525 which arrived from the North Yorkshire Moors Railway and awaits restoration to full running order.

Large Locomotives of the Great Central (1976) PLC

As its name suggests the Main Line Steam Trust, the Great Central's charitable supporting body and original administration organisation, aims to preserve main line steam. Unfortunately the scheme to restore the double track main line between Leicester Abbey Lane Sidings and Nottingham did not come about but the other aim of the Trust is more than fulfilled. The latter is a unique feature of the preserved line as it has engines representing all four of the pre-Nationalisation companies in addition to one from the pre-grouping era – this being No.506 *Butler Henderson*.

The pre-Nationalisation representatives are as follows: from the London & North Eastern Railway, No.1306 *Mayflower* and No.1264; from the London Midland & Scottish Railway Nos 5231 and 8305; from the Great Western Railway No.6990 *Witherslack Hall* and No.5224; from the Southern No.34039 *Boscastle* and No.35025 *Brocklebank Line*. In addition to these there are two fine examples of the British Railways Standard design engines, the first being No.71000 *Duke of Gloucester* which was the only one of its type built and No.92212, one of the last batch of steam locomotives built by BR. Also described in this section are the two diesels which have performed well on the Great Central so far, namely No.40 106 and the Class 127 DMU.

No.506 *Butler Henderson*

The only surviving passenger locomotive from the original Great Central Railway, it is a member of the 11F ('Director') class 4-4-0s (LNER Class D11) built and designed at Gorton works in 1919 by John G. Robinson, the Chief Mechanical Engineer. No.506 is no stranger to Loughborough Central as it used to work through the station right into British Railways' days until the 1950s when withdrawn from service. After this No.506 was set aside for preservation as part of the National Collection and restored to her former condition.

Butler Henderson's first home after restoration was the British Transport Museum at Clapham which was later moved to York. During the move of rolling stock to York however, arrangements were made with British Railways for the locomotive to be housed at Loughborough as a static exhibit, arriving there on 14th March 1975. In time, growing opinion pressed the Great Central Company to invest in a programme of works to bring the locomotive back to working order. Much work had to be done on the firebox and the boiler which took some four years and was undertaken by Babcock Power Ltd. After much effort and painstaking attention No.506 was finally steam tested on 25th March 1982 prior to entering service on Easter weekend five days later.

No.1306 *Mayflower* and No.1264

The first of these two engines, of the LNER B1 class 4-6-0s, No.1306 arrived at Loughborough in 1978 having being bought from Steamtown Carnforth from where it had worked enthusiasts' specials over British Rail's tracks. Whilst at Carnforth the engine had been given the number 1306 and named *Mayflower*, as originally carried by No.61379. No.1306 was introduced in 1948 as British Railways No.61306 and spent most of its life hauling various trains from Kingston upon Hull, although it is known to have worked in the Marylebone area on a number of occasions. After withdrawal from British Railways service in 1967 it was purchased privately for preservation and repainted in its present livery of LNER apple green.

Unlike No.1264 this engine was in first class external and mechanical condition and required only

GCR 11F Class 4–4–0 No.506* Butler Henderson, *all ready for the Easter '86 Road Rail Steam event, waits to back down on to the 11.45 departure from Loughborough Central on 31st March 1986. ***(Author)***

L N E R
4744
L N E R
1306
1306

61306

Left: The N2 class 0–6–2T couples up to the newly arrived B1, No.1306 at Quorn & Woodhouse, 18th July 1978, ready to haul it to Loughborough. *(M. Tye)*

Left, bottom: No.61306 in BR days on Great Central line metals, just outside of Marylebone station, 30th August 1966. *(P. D. Nicholson)*

retubing and boiler maintenance before entering service on the Great Central Railway. No.1306 has proved very popular over the years though in Spring 1984 the engine was taken out of service for a number of repairs prior to its trip to the Birmingham Railway Museum, Tyseley in mid June of that year. Whilst there the locomotive was steamed on thirteen occasions being placed on the depot's wheeldrop where repairs to the rear driving wheels and axles took place. It had been expected to return to Loughborough at the beginning of 1985 but this did not happen due to the ashpan needing attention. Eventually *Mayflower* returned to Loughborough in November 1985 but in a non operational state. It was returned to service in June 1986 although due to be withdrawn for major boiler work six months later.

Sister locomotive, No.1264 is the only other B1 locomotive in preservation, though much work remains to be done in order to return it to working order. No.1264 is a fine example of Edward Thompson's standard general utility engines built to replace a number of obsolete and uneconomical locomotives all over the London & North Eastern Railway network, which included the London Extension. No.1264 was known to have worked over the Great Central in the last years of its life up until its withdrawal from Colwick shed in 1965. It was then transferred to Departmental stock as No.29, serving as a stationary boiler at Colwick until July 1967. During its working life it hauled many top link runs including the famous boat trains such as 'The Hook Continental' and 'The Scandinavian'.

No.1264 was purchased from Woodham

In 1984 **Mayflower** *went on loan to the Birmingham Railway Museum, Tyseley. It is seen here on 'The Santa Special', 2nd December 1984.* *(P. D. Nicholson)*

The only surviving B1 class 4–6–0 dating back to LNER days is No.61264 built in 1947. Depicted at Barry 29th June 1976 prior to departure for Loughborough. *(R. Hardingham)*

Brothers, Barry scrapyard in South Wales by the Thompson B1 Locomotive Trust in 1975 and moved to Loughborough the following year. Unfortunately the locomotive was at one time considered beyond repair, though restoration continued. The boiler was the main cause of concern and it was even considered the locomotive would only be used for spares. However this boiler work is now being carried out.

No.61264 under restoration at Loughborough in 1986. Behind the makeshift shed, right is the Fowler diesel **Arthur Wright.** *(D. Wilcock/GCR)*

No.5231 *3rd (Volunteer) Battalion Worcestershire and Sherwood Foresters Regiment*

At present No.5231, the first of the British main line engines to operate over the preserved section from Loughborough to Quorn & Woodhouse now lies in a dismantled state in the shed at Loughborough. It undergoing a complete overhaul after withdrawal from service in spring 1977. It is not known when the engine will return to service but in the past, especially in the early days of the Railway's share issue in 1976, the engine featured largely in the press and on television during the promotion of the Great Central Company.

By 1976 No.5231 was celebrating its 40th birthday having been built in 1936 by Sir W. G. Armstrong Whitworth and Co. Ltd. of Newcastle upon Tyne. No.5231 was one of no less than 842 of one of the best known classes of locomotive in Britain, Sir William A. Stanier's Class 5 4-6-0 which was introduced on the London Midland & Scottish Railway. No.5231 worked from a number of depots on the former LMS system before being retired at Carnforth in August 1968 as No.45231, being one of the final steam locomotives to run on the British Railways' network. After withdrawal she was purchased by her present owner and moved to Loughborough Central in 1973. Three years later, on Sunday 9th May 1976 No.5231 was ceremoniously named *3rd (Volunteer) Battalion the Worcestershire and Sherwood Foresters Regiment* by her Majesty's Lord Lieutenant of Leicestershire at Quorn & Woodhouse in the presence of a local contingent of the famous TAVR unit.

For the first few years of operation No.5231 performed the mainstay of services. On some occasions the engine carried the headboard 'The South Yorkshireman' formerly the title of a British Railways' train which ran on the London Extension through Loughborough.

No. 48305

One of two recent arrivals to the railway, this locomotive is another Stanier LMS design known as the 8F class which had a 2-8-0 wheel arrangement. From their introduction in 1935, a total of 665 were built. The one preserved on the Great Central No.48305 was delivered on a low loader to Quorn & Woodhouse from Barry scrapyard in South Wales on 20th November 1985. At the time No.48305 was acquired it came with no tender but one will be purchased in the near future. After a couple of months the 8F was towed to Loughborough and shunted on to a piece of newly laid track behind the locomotive shed. As soon as this had been achieved dismantling began in earnest with the boiler being lifted on 14th April 1986.

No. 35025 *Brocklebank Line*

The second engine to be rescued in recent years from Barry scrapyard was No.35025 *Brocklebank Line* which stood for months nose to nose with No.48305. Agreement was finally reached between the Great Central Railway and the owners of the rebuilt 'Mer-

Rebuilt 'Merchant Navy' class 4-6-2 No.35025 **Brocklebank Line**, *in 'as arrived' condition at Quorn & Woodhouse, February 1987.* *(P. D. Nicholson)*

chant Navy' class 4-6-2 enabling the engine to come to the railway in Spring 1986. Like No.48305 only the engine has arrived as the original tender was previously purchased for use on the Mid-Hants Railway, so an alternative will have to be found. The owners may have to emulate another group who are having a Bulleid like tender built on the bogie of a Class 40 diesel locomotive. The latter would prove costly but may provide a solution to the problem, though because of financial constraints the final restoration of No.35025 to running order is expected to be protracted. On account of this it will be based at Quorn & Woodhouse where it will be painted to a high standard and serve in the interim as an additional attraction at the station. The last time a 'Merchant Navy' ran on the Great Central was on 3rd September 1966 when No.35030 *Elder Dempster Lines* ran a last day special from Waterloo to Sheffield and back.

No.34039 *Boscastle*

Possibly the finest locomotive on the Great Central, apart from *Duke of Gloucester*, 'West Country' class No.21C139, later No.34039 in British Railways days, was built at Brighton Works in 1946 to Mr O. V. S. Bulleid's controversial design. It was streamlined until 1959 when modified to her present form. During her working life with British Railways it frequently worked between Waterloo and Bournemouth with heavy express passenger trains. Finally withdrawn from service in 1965 it joined many other redundant steam locomotives at Woodham's scrapyard, Barry where it lay until rescued by James Tawse, one of the Main Line Steam Trust directors. *Boscastle* arrived at Loughborough in January 1973 since which time heavy restoration work has been in progress. It was originally estimated that the engine would only take three or four years to put back into working order but certain over-riding events have hampered efforts, one of the latter being the lack of funds to complete the project. This has resulted in the ownership being transferred to a syndicate. (At one point even the 'unthinkable' was considered, namely moving this magnificent locomotive to another preserved railway!) Along with *Duke of Gloucester*, *Boscastle* has one of the most interesting stories as far as restoration is concerned.

No.6990 *Witherslack Hall*

Owned by the Witherslack Hall Locomotive Society, it arrived at Loughborough in 1975 and was under restoration in the shed there for the next eleven years until finally steamed on a test run in

No.6990 **Witherslack Hall** *made a very welcome return to the Great Central line in 1986. Seen here on 19th October in fine form.* *(D. Wilcock/GCR)*

the spring of 1986. The locomotive finally entered service on passenger trains in the autumn of the same year, resplendent in Great Western Railway Brunswick green livery. No.6990 was built at Swindon in 1948 as one of the 'Modified Hall' class of engines incorporating design improvements made by F. W. Hawksworth increasing the performance and reliability of C. B. Collett's original 'Hall' class introduced in 1928.

Witherslack Hall is no stranger to the Great Central as along with many other 'Halls' it used to work over the Great Central London Extension, hauling express cross-country trains via the Banbury to Woodford Halse link which connected the Great Central to the Great Western. No.6990 also ran on the Great Central in the locomotive exchange trials of 1948 competing with an LMS Class 5 4-6-0, an LNER B1 class 4-6-0 and a SR 'West Country' class 4-6-2, which are all represented on the present day Great Central Railway. The trials saw some unusual mixtures of engines and tenders which was all done to see if locomotives were capable of running on lines other than their own with different weight loads etc. This was also the beginning of steps by the newly formed British Railways to design a number of standard locomotives that would be capable of being moved from region to region and still give the best performance. To measure their performances during the trials engines like No.6990 were coupled to a Dynamometer car.

No.71000 *Duke of Gloucester*

This locomotive was built in 1954 as the prototype of a new standard design of engine for hauling main line expresses, the Class 8P 4-6-2. However because of the then Modernisation Plan of the British Transport Commission the following year, the decision to build further steam locomotives was shelved in favour of constructing diesel locomotives. Therefore No.71000 *Duke of Gloucester* was by a quirk of fate to become the one and only member of its class. Following its withdrawal from service in 1962 it was scheduled for preservation with many other interesting locomotives.

Change in policy meant that the 'Duke' lay in store at Crewe North locomotive shed until finally it was removed to the scrapyard at Barry in 1967 for breaking. In 1973 a group known as the 71000 Duke of Gloucester Locomotive Trust was established to rescue the locomotive and restore it to running order. The efforts of the Trust were to pay off although from the outset the project of restoration was to be a daunting one. In 1967, shortly before it was moved from Crewe the cylinders and Caprotti valve motion were removed for display in the Science Museum and the tender was sold for use in a steel works as an ingot.

Final purchase of the remnants of the engine were made from Woodham Brothers and on 27th April 1974 the engine was transferred to Quorn & Woodhouse by road and from there it was towed by rail to Loughborough. On arrival the engine began to be stripped down and the firebox, boiler and smokebox were lifted, along with the cab to allow attention to the driving wheels and main frame to be carried out. While this work was being completed new drawings of the valve gear were being draughted with a view to new sets being fabricated.

With the arrival of larger engines such as No.71000 **Duke of Gloucester** *and the need for provision of a storage area to shelter and restore locomotives from the elements, a running shed was constructed between 1973 and 1974. This was carried out entirely by the volunteer work force, and is seen here taking shape over 'The Duke'.* *(Author)*

Several simple tasks were performed which included the red oxiding of parts and the steam cleaning of the main chassis. At the same time an appeal was set up to raise funds for the valve gear. By 1977 the cam boxes for the valve gear were being made and progress made with the cylinder casting patterns. Much work was done on the driving wheels and the pony truck which was reunited with the main chasis in August 1977. Other work at that time was effected on the tender which originated from 9F No.92134 at Barry and which was treated to three new wheel sets.

The following spring the boiler was put back onto the 'Duke' which allowed a number of other fittings to be installed. At the same time the boiler tubes were removed and repairs were carried out on the flues which were re-ended and replaced. In addition the ashpan was completed and refitted while work was beginning on the manufacture of the cylinders. Valves were being made by Froude Engineering in Worcester, while a £5,000 grant was obtained from the Science Museum in London to help with the replacement of the cylinders and motion. In general the locomotive was now beginning to take shape. The next stage saw the completion of the tender and its coupling to the engine.

The next six years saw the complete remaking of the blastpipe and subsequent refitting, the retubing of the boiler, the machining of the cylinders and their installation and the overhaul of the superheater. In spring 1983 the tender and cab were painted for the first time in green undercoat and pipework was completed for the sanding gear, after which work commenced on the main pipework. At about this stage the superheater elements were hydraulically tested. Aside from the main pipeworking and fittings for the cab, the front and rear ashpan dampers were installed. The main task now was the acquiring of sufficient funds to have the coupling rods machined. While funds were being raised for this, various other tasks were carried out such as the hydraulic testing of the boiler. By autumn 1984 following a successful appeal through 'Steam Railway' magazine, which raised some £3,800, all the coupling and connecting rod forgings had been purchased and sent for machining. Following this the pistons were returned and assembled in the cylinders which in turn were fitted to the crossheads prior to the cylinder coverings being attached. Various other hydraulic tests were made to the boiler after which the chimney was bolted down, the superheater elements and header were fitted, the cladding for the boiler and the numerous castings and unions for the pipework attached.

The outside cylinders and motion on the newly restored **Duke of Gloucester** *which had to be completely made anew for both sides of the engine.* *(Author)*

On the cosmetic side painting of the cab was started together with the tender and other items, ready for the expected return to steam in 1985. In the event the naming ceremony and official commissioning did not take place until late 1986 for a number of reasons, but, I think all who see it will agree it was well worth the wait.

A remarkable achievement by all those involved and a credit to the Railway as a whole – the fully restored BR Standard 8P class 4–6–2 No.71000 **Duke of Gloucester,** *at Loughborough Central in late 1986.* *(Author)*

No.92212

In the same year as *Duke of Gloucester* was introduced, the 9F class 2-10-0s appeared which were destined to be the final steam locomotives built by British Railways. Designed to be able to haul heavy freight trains, a total of 251 engines were built at Crewe and Swindon works between 1954–60. The very last steam locomotive No.92220 *Evening Star* was commissioned in March 1960 and now survives as a part of the National Collection.

Although mainly a freight engine, the 9Fs had an extremely successful service record and when called upon to work passenger trains they achieved some remarkable performances. Many of the class, including No.92212 worked over the Great Central London Extension in the early 1960s. They were especially noted for their performances on the Annesley to Woodford fitted freights, the 'Runners' and 'Windcutters'.

After her withdrawal in 1968 No.92212 was despatched for scrapping with eight other engines of the same class to Barry scrapyard in Wales. The engine lay there for ten years until 1978 when she was rescued, a previous attempt having failed. By 1978 the engine was nothing short of 120 tons of rusting steel almost 70ft long-devoid of all non ferrous fittings, and connecting rods. Ten years of inactivity had seen a dreadful decay set in.

The locomotive was moved to Loughborough in two stages with the tender being transported in November 1978 which was the last of its type left at Barry. The locomotive was moved in September 1979 and deposited at Quorn & Woodhouse where it remained until it could be moved to a space at the back of the locomotive shed at Loughborough. With the employment of twelve adults on a Manpower Services Commission scheme, work on the tender progressed rapidly and was completed and liveried in British Railways' black.

The ex Barry 9F, No.92212 at Loughborough in May 1980 prior to the erection of its shed, which has more or less kept it from photographers whilst the restoration has been progressed.
(P. D. Nicholson)

Following the completion of the tender, the locomotive was dismantled and a shelter was built over it in preparation for the grit blasting of the boiler, cab, frames, cylinders and wheels. Shortly after this the front bogie was fully restored and the driving wheels were removed to check the bearings. At this juncture attention was turned to the tubes and flues which were removed from the boiler. By spring 1983 the smokebox was removed for replacement and a new ashpan fabricated, at the same time as the re-assembly of the motion was being completed.

During 1984 much work was undertaken at Swindon, on the driving wheels, journals and crankpins as well as on a seized brake arm. In addition other areas were being attended to such as the ashpan which was re-assembled, the bufferbeam, the front tubeplate, the firebox, axleboxes and lubrication system. By the autumn of the same year work on the boiler cladding was largely completed and the copper piping for the lubrication system had been acquired. Completion was also being achieved on the driving wheels, frame and cab in respect of painting. Over the next two years many small tasks were undertaken on the axleboxes, front bogie and buffer beam. More intricate work included the fitting of copper piping and the assembly of the running boards which have been remade. Though work progresses apace a date for the engine's steaming has yet to be fixed.

No.5224

An example of Collett's 5205 class 2-8-0 tank which was a development of a 1910 Churchward design intended as a powerful engine for hauling heavy mineral trains over comparatively short distances. No.5224 spent most of its working life in South Wales and was one of 205 of its type having been built at Swindon in 1924. One of the advantages when it came to preservation was the fact that its strong and robust design meant that it was in a favourable condition when acquired by its new owner Mr R. Hibbert in 1978. Since then the locomotive has undergone a heavy restoration programme and was decked out in a British Railways' black freight livery before being steamed in 1985.

No.5224 basks in the sun on platform 1 in front of various coaches awaiting restoration. This locomotive was first steamed in 1984 and has been a regular performer ever since. In this picture it was seen prior to working an afternoon train to Rothley. (Author)

No.4744

GNR N2 class 0-6-2T No.4744 (BR No.69523) was built by the North British Locomotive Co. in 1922 to the design of Sir Nigel Gresley. These locomotives chiefly worked the heavier suburban trains in the King's Cross and Moorgate area. Locomotives of this class were originally fitted with feed water pumps but these were removed when injectors capable of dealing with water heated by the condensing apparatus (necessary for the tunnels to Moorgate) became available.

No.4744 saw extensive use on the Keighley & Worth Valley Railway during which time it featured on the 'Scotch Express' in the film 'The Railway Children'. In 1975 No.4744 was moved to the Great Central Railway and after an extensive boiler overhaul at Loughborough it returned to service in 1978. It has since been a regular performer, operating under a special licence from its owners, the Gresley Society.

The N2 No.4744 approaches Loughborough Central 3rd May 1980. (P. D. Nicholson)

Robert Stephenson & Hawthorns 7597 of 1949, 0–6–0T at work at its former location, CEGB Rye House Power Station, Hoddesdon, Herts, 18th November 1967. (P. D. Nicholson)

No.7597

Although a former industrial locomotive No.7597 has been used on the heaviest trains. It arrived on the Great Central on the 10th May 1982 from the Stour Valley Railway Preservation Society site at Chappel and Wakes Colne, Essex. It is an 0-6-0 tank built in 1949 by Robert Stephenson and Hawthorns Ltd for use at Central Electricity Generating Board, Rye House Power Station, Broxbourne, Hertfordshire. No.7597 was used mainly as a back-up engine to sister locomotive No.7598 (scrapped in 1962) from which time, until 1971 it was employed on shunting coal wagons around the plant. After this it was acquired by Railway Vehicle Preservations Ltd who transported it to the Stour Valley Railway Preservation Society HQ where she was a regular performer until 1976. Following some minor repairs No.7597 became a leading light in train operation for a number of years after its arrival at Loughborough, though at present it is out of service pending a ten year boiler examination.

No.377 *King Haakon 7*

This former Norwegian State Railways Class 21c, 2-6-0 was built in Sweden (Nydqvist & Holm 1164 of 1919). It was acquired for preservation and imported to the UK in 1969, going first to the GCR at Loughborough but is now at the Bressingham Steam Museum, Norfolk.

Class 21c 2–6–0 **King Haakon 7**, *formally of Norwegian state Railways, in steam at the southern end of Loughborough Central station on 29th July 1973, in the early years of brakevan rides and main line operation. (M. Hooper-Immins)*

No.40 106 *Atlantic Conveyor*

The Great Central's only large, main line diesel locomotive, it arrived at Loughborough shortly before Easter 1984. It had been stored out of use at Crewe for 13 months and was purchased by its present owner as it stood. The locomotive had a rather complex journey to get to the GCR, but once it had arrived restoration work was started immediately, with battery cleaning, topping up and charging. Following this the engine was barred over to check for obstructions in the cylinders. Water was found in three cylinders and released appropriately. Next the cooling system was flooded with water to check for leaks though none was found. Attention then turned to the electrical system to allow the engine starting to take place and after such fuel pressure was attained and four attempts at turning the engine, the 'Whistler' roared into life and the locomotive was able to move under its own power.

After a successful starting, various minor repairs were carried out on the brakes, traction motors and the bogies were checked. Also a complete lubrication was undertaken on the underneath. Once all minor repairs had been carried out, the yellow nose ends were repainted as were the bogie frames and buffer beams. A major service was effected on the power unit and No.40 106 returned to service sporting the green livery she had worn when new although retaining its BR TOPS identity, rather than its original number, D 306. No.40 106 was one of 200 English Electric locomotives built at Newlon le Willows and Darlington.

Once working again, on the 9th August 1984 No.40 106 was named *Atlantic Conveyor* in memory of the ship of that name lost in the Falkland conflict.

***Not only does the Great Central preserve steam locomotives but also diesel locomotives and here, fully restored to British Railways green livery and carrying the headboard 'The Green Goddess' is No.40 106. It is seen drawing into Loughborough from Rothley with Dave Hartley at the controls. The locomotive is better known now as* Atlantic Conveyor, *named in honour of the ship destroyed in the Falklands campaign, on 9th August 1984.* (Author)**

Diesel Multiple Unit

This unit comprises of two Class 127 power cars Nos M51622 and M51616 plus a Class 120 buffet trailer car No. M59276 which were delivered to Quorn on 2nd March. Various minor works were carried out prior to it running on driver training

runs. When acquired there were no corridor connections, so the Great Central has installed them. The unit now sees regular service and looks very pleasing in a version of the old dmu green livery.

Class 127 diesel multiple unit, formerly used on St. Pancras – Bedford services. The centre vehicle, which provides the buffet was acquired from another unit and sandwiched in the middle of the driving trailers. A new feature for the DMU has been the installation of corridors and the unit is a regular performer on the 10.45 service. Car No.M51622 nearest the camera, February 1987. *(P. D. Nicholson)*

tion of the preservation of two steam engines and two of the diesels. A list of all the other small engines that have been or still are resident on the Loughborough to Rothley line is appended. The first engine is a former NCB Austerity class 0-6-0ST No. 9 (Hunslet 3825 of 1954), which arrived at Quorn & Woodhouse on 10th November 1981 having been purchased by the Earl of Lanesborough and donated to the Railway. After being towed up to Loughborough work began almost immediately on stripping it of its fittings in preparation of a

Small Locomotives of the Great Central Railway (1976)

The Great Central is fortunate in having a large collection of small industrial steam and diesel locomotives acquired from both industry, and British Rail and its constituents. Some of the more powerful members of the latter category are employed on main line services while the others are confined to shunting or inhabit the site in an unrestored condition.

To give an idea of how long the restoration of an industrial steam and diesel locomotive can take and what it involves this section includes a description of the preservation of two steam engines and two of the diesels. general overhaul and painting programme. Included in this work was the retubing of the boiler which had not been used for three years. By the Spring of 1982 the tank, cab and boiler were all removed and the buffer beam was taken off for straightening and the tubes from the boiler were extracted.

Once the engine was moved inside the shed, further work included the dismantling of the connecting rods, valve gears and wheels which were subsequently turned. Following the turning of the wheels the axlebox underkeep pads were sent away to York for attention. This was closely followed by the replacement of the wheels. The next task was to repair the boiler which was undertaken by Babcock

Power Ltd and while this was in progress, a new bunker was being fabricated at a local skill centre. The latter was completed in June 1984, as was the boilerwork and assembled by apprentices on work experience from the BREL carriage and wagon shops at Derby. Following the retubing and attention to other minor details, the locomotive was outshopped in an unlined black livery carrying a British Railways' style number 68009 and lettering, being similar to the LNER/BR J94 class. It was not

The Austerity (Hunslet 3825) on shed, February 1987. Despite its number, 68009 and lettering this locomotive does not originate from LNER/BR stock and has therefore never been classified as a 'J94'. ***(P. D. Nicholson)***

steamed until May 1986 after repairs had been carried out to the firebox.

The second locomotive was one of the London & North Eastern Railway Y7 class No.985. It was built as North Eastern Railway Class H8 to the design of T. W. Worsdell as a chunky little engine weighing a mere 22tons 14cwt. An 0-4-0 tank engine, No.985 arrived at Loughborough Central towards the beginning of 1981 where it was stripped down for overhaul. Following this a brand new boiler, which came with the locomotive, was fitted and work progressed steadily. As time went on several people were delegated to the rebuilding of this small locomotive, who completed the cleaning of the motion gear and several other tasks which needed doing. Later, in 1983 the boiler was hydraulically tested and the valve gear was re-fitted along with the chimney, cladding, side tanks and cab. By October the Y7 had been painted in unlined black with British Railways lettering and number, 68088 and was duly steamed. It was taken out of service during the winter in which it was first steamed and then in the following year the locomotive became a part of the Great Central publicity machine. The Y7 was loaded onto a low-loader and taken round the streets of Nottingham with some of the Railway's staff as part of the celebrations in connection with the Nottingham Annual Festival. This has continued to be done for two years now but concern was mounted over the fact that the engine seemed to be covering more mileage by road than rail. In an effort to remedy this situation the Y7 now provides short rides between the shed at Loughborough and the station platform on special days and Bank Holidays.

In addition to these engines the Great Central has many other former industrial steam locomotives which are listed below, though because they are not owned by the railway company some have moved to other sites over the years.

Small Steam Locomotives

No.39

This locomotive (Stephenson & Hawthorns 6947 of 1938), was converted to a side tank design by members of the Railway from a saddle tank. This enabled it to haul passenger trains, and being turned out in Great Central passenger livery the engine was a special attraction. High passenger loadings ensured this 0-6-0 saw a fair degree of use but by May 1980 the Leander Locomotive Group decided to transfer No.39 to Carnforth where it has since been put up for sale.

No.39 (Stephenson & Hawthorns 6947 of 1938) the former 0–6–0 saddle tank, rebuilt with side tanks by Mr Roger Hibbert and painted in lined green GCR passenger locomotive livery. Photographed at Loughborough in May 1979 it was later moved to Steamtown, Carnforth by its owners, The Leander Locomotive Ltd. *(M. Tye)*

Littleton No.5

Manning Wardle 2018 of 1922, 0-6-0ST, also described in Chapter 5 was a regular performer up to March 1980 when the locomotive was advertised for sale in the railway press. The proceeds of the sale which eventually went through in the autumn of the same year were put towards the restoration of GWR No.5224. *Littleton No. 5* was transferred to the Bitton Railway near Bristol.

Lamport No.3

Built by W. G. Bagnall in 1942 as works No.2670, this engine was supplied by the Ministry of Supply to Scaldwell and Hanging Houghton Pits in Northamptonshire, from which ironstone was extracted. The locomotive was fitted with automatic couplings for use with steel dump cars. It was acquired in 1969 by its present owners, exhibited at the Lord Mayor's show in 1970 and subsequently moved to Loughborough in April 1973 when it was painted in a Great Central green and lined out in black and white. The engine now resides on the Market Bosworth Light Railway.

Marston's No.3

Hawthorn Leslie 3581, built in 1924 it carried a livery of Mediterranean blue. This locomotive, which was a stalwart performer in the early days of the Great Central now lies dismantled awaiting restoration.

Hilda

Built in October 1938 by Peckett, their number 1963

it was acquired by Mr R. Bailey and Mr D. Kemp from British Gypsum at Newark and then transferred to Loughborough. *Hilda* is also still lying on the site in a dismantled state awaiting restoration.

Robert Nelson No.4

Hunslet 1800 of 1936 this 0-6-0ST is still a regular performer on the railway on passenger trains. To commemorate the locomotive's 50th anniversary it has been turned out in a lined GCR black livery and a small celebration was held on 15th June 1986 to mark the event.

of 1975 while *St Monans* (Sentinel 9373 of 1947) remained inoperative. In power these engines were quite strong though small and in appearance were closer to the design of a diesel rather than that of a steam locomotive.

As well as steam locomotives the Great Central also has a fair selection of small diesels, the largest of these are, namely the ex BR Class 08, 10 and 14s. The first two are both owned by Mr Tony Naylor who purchased the Class 10 Blackstone diesel electric 0-6-0 shunter from the Nailstone Coal Preparation Plant near Coalville. Although the

Robert Nelson No.4 ***after repainting in Great Central lined black livery for an anniverary special train, which ran on the 15th June 1986 to celebrate the locomotive's Golden Jubilee. No.4 has been the mainstay of winter services and is a stalwart performer on the Great Central.*** ***(Author)***

locomotive had a new home to go to, at the time it was unable to be loaded onto the road vehicle waiting to take it away to Quorn & Woodhouse. The reason for this was the engine was locked in gear. It was only possible to move it after extra winching gear had been brought in from Northampton. However shifting the locomotive onto the trailer was not going to be so easy as the volunteers thought as a pin then sheered as it moved forward. D4067 as the locomotive is better known, was one of 146 built of this version of the ubiquitous 0-6-0 diesel shunter, many eventually being owned by the

Other locomotives which were at Loughborough in the early days included two Sentinels steam locomotives *St Monans* and *Neepsend* No.1 which came from Newark on Trent and a Sheffield power station respectively. *Neepsend* No.1 (Sentinel 9370 of 1947), which was painted in green livery operated frequently on trains during the summer

National Coal Board. Having arrived at Loughborough in February 1980, work began on the cleaning of electrical equipment and the removal of one of the traction motors which had been burnt out. In Summer 1982 one of the traction motors was rebuilt and replaced having had the gearbox on this motor attended to. By 1983 the Class 10's number two traction motor was repaired and ready for refitting, after which, work started on the repainting of the bodywork in British Railways livery. Following this the main control panel was checked over and the gearbox on traction motor two was rebuilt. young, making the locomotive unavailable for service! This engine is now used quite often on stock movements and has proved a valuable asset. A similar locomotive was bought in December 1984 from Amey Roadstone at Loughborough in the shape of Class 08 No.3101 which is to be preserved by Tony Naylor in British Railways' black livery and perform the same duties as D4067.

The next type to be described is the Class 14 650hp, 0-6-0 diesel hydraulic, three of which are based at Loughborough. These work main line train services during the winter and on special days in the

Class 10 diesel electric 0–6–0 shunter No.D4067 which is mainly used for works trains, seen here about to start its next turn of duty on 14th July 1986. ***(Author)***

Other jobs at the time included a hydraulic test on the air tanks which were found to be serviceable. In Spring 1984 both motors were tested and minor adjustments had to be made to the gearbox on traction motor one before the locomotive was fit to enter service on crew training runs. During the crew training period a blackbird decided that the frames of the shunter would make an ideal place for her

summer. The first two arrived at Loughborough on 16th October 1981 acquired from British Steel Corporation, Corby. Numbered (BR) D9523 (BSC) 46 and (BR) D9516 (BSC) 56, they were both professionally serviced during the winter 1981/2 ready for use on main line trains. Diesels are regarded as an important asset to the Great Central as far as running costs are concerned. They are also considered a benefit as they attract diesel enthusiasts to the railway. No.D9523 was built at Swindon in 1964 and sent to Old Oak Common in December of that year. From there it was sent to Bristol Bath Road until October 1965 before finally being transferred to Hull in January 1967 prior to its withdrawal in April 1968. It was then sold to Stewarts and Lloyds Ltd, Corby in December 1968. Originally numbered by them as 8311/25, later becoming No.46 it was allocated to Glendon. No.D9516 was similarly built at Swindon in 1964 and sent to Cardiff Canton depot in the October. It then moved to Landore in May 1965 and finally to Hull in 1966 before being

One of three ex BR Class 14 0–6–0 diesel hydraulics, No.D9516 at Loughborough in 1987. *(P. D. Nicholson)*

A busy scene at Quorn & Woodhouse as Class 14 D9523 draws into the station with a train from Loughborough Central. Note the barley sugar lamp posts and the typical layout of a Great Central island platform. On the right is the goods yard which now sees most use at Easter on the Road Rail Steam days. *(D. Bonas)*

sold to Stewarts and Lloyds, Corby in November 1968. Originally designated 8311/36 then No. 56, the locomotive was then allocated to Corby.

These have proved an ideal type of engine for the five and a half miles of line run by the Great Central. A further member of the class was added to stock on 11th December 1984 when No.D9529 was acquired from the North Yorkshire Moors Railway. At that time the locomotive was non operational due to having a sheared auxiliary generator drive shaft inside the Voith transmission. Work started on this as soon as the locomotive arrived and the transmission was assembled and the engine was in running order by the end of 1985. On the 29th September that year it appeared resplendent in its new rail blue livery complete with British Rail style TOPS number 14 029. The locomotive never carried the number or the blue livery in its British Railways' life, but was so liveried by the Great Central in order to provide Class 14 enthusiasts with an idea of what they would have looked like had the engines survived in BR service past 1968.

In addition to the diesels aforementioned there are numerous other types in existence on the Great Central, some of which are described below.

Small Diesels

D4279

Fowler 4210079 of 1952, an 0-4-0 diesel arrived on the Great Central in January 1975 and performed stalwartly on various trains when no steam locomotive was available. The locomotive was also used extensively on shunting operations in the days before the passing loop was installed at Rothley. Other uses of this small unit included shunting and pilot duties at Loughborough and it helped in running the railway at less cost, a factor which was all important at that time. In August 1975 the diesel was named *Arthur Wright* after one of the volunteers who had recently died.

By 1977 deterioration of this locomotive had occurred, so it was put on a limited availability programme to ease the wear being put on it. Temporary repairs were made to the radiator but the locomotive had to be taken out of service for two months in the autumn of 1977 for major attention to the gearbox and drive system. Additional pipework for the air horns had to be installed and once all this work was finished new brass nameplates were fitted to the cabs to commemorate D4279's return to traffic. During the next ten years up until 1981 *Arthur Wright* saw limited use before being withdrawn again from service.

Fowler 4210079 of 1952 0–4–0 diesel No.D4279 **Arthur Wright** ***sits at the back of the shed at Loughborough awaiting restoration. The locomotive was a stalwart performer in the early days of the railway and was named after a MLST director who died prematurely. (Author)***

QWAG

A small 4-wheel Ruston & Hornsby diesel shunter, 371971 of 1954 which was originally owned by Frederick Parker of Leicester and based at their Viaduct Works from 1954 to 1965 when the company's private sidings were severed from the British Railways' system. The locomotive was acquired from the directors of Parker Plant by members of

The diminutive 48hp 4-wheel diesel, No.1 **QWAG** ***(Ruston & Hornsby 371971 of 1954) at Loughborough in July 1980. (M. Tye)***

the Quorn & Woodhouse Action Group from whence the name *QWAG* is derived. After arrival in December 1972, a complete mechanical and external restoration was carried out and the locomotive was decked out in a blue livery prior to being put back into service. *QWAG* was to prove a useful asset in shunting operations in the early years along with *Arthur Wright*.

No.11 *Alen Grice*

Originally built as a Sentinel 4-wheel geared steam locomotive, in 1963 it was rebuilt a year later by Thomas Hill & Sons to diesel power, 134c of 1964. Incorporated in this rebuild was a high performance turbo charged Rolls Royce diesel engine, an improvement on its steam equivalent. In 1980 the locomotive was purchased in working order and transferred to Loughborough. Used mainly for shunting in the yard at Loughborough and occasionally on engineers' trains, it was repainted in 1981 and named *Alen Grice* in recognition of the splendid achievements of the Chief Locomotive fitter of the Railway.

Following a long period of trouble free running, on 11th March 1985 No.11 was derailed on the crossover north of Loughborough Central. Sadly in this mishap one of the axles cracked and the disabled locomotive had to be lifted clear of the rails by the Great Central's steam crane. When the axle was removed for insurance assessment, it was realised that the cover payment would not be sufficient to obtain a new axle, so a secondhand one would have to be found and until that time the locomotive would have to remain disabled. Spring 1986 saw work start on No.11, once the required wheelsets had been successfully purchased. At the same time an ultrasonic test was made on all the axles which were found to be in the best of order. While this was being done the trailing wheelset was sent away to Morcroft Wagon Works for re-profiling before refitting.

No.28

This 0-4-0 diesel was offered as a donation to the Great Central Railway by Bardon Hill Quarries and it arrived at Loughborough Central on 18th June 1985, a welcome addition to the shunter stud, especially as No.11 *Alen Grice* then lay out of action. No.28 last operated regularly in 1981 and had been sporadically used until disposed of by the quarry company. A locomotive of small size, No.28 built by Andrew Barclay & Sons & Co. at the Caledonian Works in Kilmarnock in 1956 (No. 400), has a maximum speed of 10mph, weight of 34tons, power rating of 204bhp and a mechanical drive with three forward and three reverse gears. Eventually this engine is due to be repainted in Bardon Hill's Brunswick green livery with yellow lining, but the name it carried at the quarry *Duke of Edinburgh* is unlikely to be carried again.

A 1964 4-wheel diesel hydraulic rebuild (Hill 134c) of a Sentinel steam locomotive. On 4th July 1981 it was found in the company of* Robert Nelson No.4. *The diesel has since been named* Alen Grice.** ***(P. D. Nicholson)

Above right: Barclay 400 of 1956 0–4–0 diesel mechanical No.28 kindly donated to the Great Central Railway by Bardon Hill Quarries Ltd, stands in front of the disabled persons coach in the siding by the old 'up' main line. (Author)

Great Central Railway coaching stock

Number	Type	Builders	Date	Notes
5664	Open Third	GCR Dukinfield	1910	'Barnum'
5666	Open Second	GCR Dukinfield	1910	'Barnum' On loan from NRM.
1852	Buffet	LNER Doncaster	1937	ex 'Flying Scotsman'.
9122	Buffet	LNER York	1937	
9124	Buffet	LNER York	1937	
16520	Corridor/ Brake Third	LNER York	1937	Repainted for use in filming work by Central Television
18033	Corridor/ Composite	LNER York	1924	(Repainted for use in filming work by Central Television)
62565	Corridor/ Brake Third	LNER York	1927	
70427	Full Brake Pigeon	LNER York	1941	
E70268E	Travelling Post Office	LNER Dukinfield	1937	
E70294E	Travelling Post Office	LNER York	1937	
W15611	Corridor/ Composite	Birmingham Carriage & Wagon	1955	
W15096	Corridor Composite	Birmingham Carriage & Wagon	1953	Disposed of.
M34393	Corridor Composite	Birmingham Carriage & Wagon	1954	
E34738	Corridor Brake Second	Birmingham Carriage & Wagon	1955	
M3042	First Open	Doncaster	1953	
E16025	Corridor Composite	BR Doncaster	1955	
E1525	Restaurant Buffet	Cravens Sheffield	1961	
E1526	Restaurant Buffet	Cravens Sheffield	1961	
E70654E	6-wheel Full Brake		1950	LNER type
S1706S	4-wheel PMV	SR Lancing	1943	

Restored 'Barnum' saloon No.5664 complete with new roof and varnished bodywork and handrails. The vehicle is used mainly for board meetings and museum exhibitions at Quorn on bank holiday weekends. *(Author)*

LNER Travelling Post Office sorting van No.E70294E, complete with post box, pick-up net and setting down equipment for transferring mail bags at speed. *(Author)*

Resplendent in carmine and cream livery the Great Central's disabled persons coach No.E9316 converted from a BSO at BREL workshops Derby. Seen at Rothley. ***(Author)***

The interior of the coach No.E9316 showing how the inside has been modified to cater for disabled passengers and able bodied travellers alike. ***(D. Bonas)***

The BR Mk 1 coaches with their classifications:

Number	Classification	Notes
380	Crew Coach	
Sc1852	Restaurant Mini Buffet	
E3079	First Open	
E3126	First Open	
E4630	Tourist Second Open	
E4662	Tourist Second Open	
S4914	Tourist Second Open	
E9316	Disabled Coach	
W15296	Disposed of	
E15514	Disposed of	
E25189	Corridor Second	
Sc24656	Corridor Second	
E25312	Corridor Second	
M27001	Brake Tourist Corridor	
E43043	Composite Lavatory	
E46139	Second Open	
E48001	Second Lavatory Open	
E57541	Brake Tourist Corridor	Sold by Gresley Society in 1986 to Railway Vehicle Preservations Ltd.
W4362	Second Open	
W4788	Second Open	
E7070	Corridor Composite	
E3095	First Open	
W18366	Corridor Second	
315	Pullman First Kitchen	*Heron*
24278	Restaurant Buffet	
24280	Restaurant Buffet	
84343	Guards Brake	

Left: Utility van PMV No.1706 in Southern Railway green livery complete with full lettering. This vehicle is mainly used on Road Rail Steam days at Quorn as a home for a model railway display.
(P. D. Nicholson)

In addition to the coaches described on page 197 the Great Central owns many other ex British Railways examples of the Mark 1 design built in the fifties/-early sixties. It would be too onerous-a-task and a monotonous one to give a complete description here of every one of these but for the benefit of those interested in coaching stock I have included a list opposite.

Coaches are in three basic liveries, blue/grey as inherited from British Railways, carmine and cream which is the standard livery of the Great Central Railway (1976) PLC and lined maroon which has recently been applied to one of the coaches that is now in service, namely No.E16025. Because of the variety of engines a change in Railway policy is about to take place with the painting of carriages in the colours of the four Regions of British Rail prior to standardisation. The latter decision will allow an engine like *Duke of Gloucester* to haul a set of carmine and cream coaches, No.40 106 to haul a rake of maroon coaches perhaps, and *Witherslack Hall* to haul a rake of GWR chocolate and cream liveried vehicles as it would have done during its original working life.

Wagons and Vans

Ballast wagon ('Catfish')
No.DB983393, 25ton capacity, 10ton tare. Built in 1956 by Metropolitan Cammell Ltd.
Ballast wagon ('Walrus')
No.DS62067 38ton capacity 22tons tare. Length 35ft. Built by the Southern Railway in 1947.
20 Ton goods brake van
LMSR, built at Derby in 1924.
14 Ton tank wagons
'Esso' built in 1941 as LMSR No.162740.
'Vickers' built March 1943 as LMSR No.164253, now used as a weed-killing wagon.
Three plank wagon
British Railways No.DM742591 12ton tare. Formerly used by the District Civil Engineer at Birmingham.
High sided wagons (3)
Formerly used by the National Coal Board for internal traffic and typical of the many private owner coal wagons once common on Britain's railways. One of these has had its sides removed and is in use as a flat wagon.

A 10 ton, 6-plank private owner coal wagon in the black and white livery of Babbington Colliery, Nottingham.
(P. D. Nicholson)

Permanent way flat wagon
Converted from former Norwegian State Railways 'Konductorvogn' 547. The teak superstructure from this coach is now located by the locomotive shed at Loughborough Central, where it serves as an office.
No.DE270809 'Dolphin'
Built by Head Wrightson in 1947, purchased for use by the Permanent Way Department.
Goods brake van
No.954268 Ex British Railways.
Ex MoD wagons
45055, 45073, 45074, 1-plank open wagons; 47528, 47534, box vans

Other vehicles
Train Heating and Generator coach (BR No.TDE321099) Rebuilt by British Rail from a standard horsebox (HB) No.96336. This vehicle is to be refurbished for the purpose of steam heating trains and providing electrical power.
45 ton steam crane (BR No.TDM 1097) Built for the Great Western Railway by Ransomes and Rapier Ltd of Ipswich in 1939. This crane is typical of a large number of steam cranes which were to be found at locomotive sheds throughout the British Railways' system.
Syphon G (BR No.W1025) Built Swindon, 1952.

Maintaining and Operating the Great Central (1976)

Although not as busy as a main line and subjected to as much wear and tear, a private railway such as the Great Central, has to maintain its permanent way and follow operating procedures on a par with the guidelines laid down by the Department of Transport. To carry out any necessary works, the Great Central possesses its own permanent way force and due to the need for signalling there is also a dedicated/skilled signal and telegraph team.

The permanent way force, since the Railway's inception has been responsible for a number of tasks such as clearing and ballasting of track. Work of this nature has been performed on the extension of the line from Rothley to Birstall requiring the removal of vegetation prior to ballast and track laying taking place. To aid these tasks the department has its own set of ballast wagons and for weedkilling the 'Vickers' tank has been fitted with a pump and chemical spraying arms to enable the liquid weedkiller to be spread over the area surrounding the track. The ballast and tank wagon vehicles are stored at Quorn. Over the years as more and more rolling stock has come to the Railway more storage space has had to be found. This has been done by the provision of extra and longer sidings at Loughborough – the work being undertaken by the permanent way team.

Extension of track was not just limited to sidings but included the installations of run-round loops like that at Loughborough Central, completed just after British Railways' contractors removed the old London Extension 'down' line in 1976. Shortly after, a similar loop was provided at Rothley, though it was not until 1985 that work was started on the Quorn & Woodhouse loop which will eventually allow more than one train to operate on the $5\frac{1}{2}$ mile line at any one given time. The building of the latter was completed within a year as a result of the outstanding efforts by the men of the permanent way team. Other works on the track include the replacement of sleepers, securing of wedges or springs and the replacement of stretches of rail from time to time.

In addition to the track laying and repairing team, a splinter group exists whose brief is to ensure that the bridges and viaduct are in good order. Jobs in this department vary from the re-pointing of brickwork or repairing of girders and sidewalls to the complete rebuilding in one case. One of the most arduous tasks was rebuilding the parapet wall at one point on the south side of the station bridge at Loughborough which was demolished by an errant road vehicle! By far the largest of the structures which come under the auspices of the bridge gang is Swithland Viaduct which crosses the reservoir

The weir viewed from Swithland Viaduct. (M. Hooper-Immins)

serving Leicester. This has been of little cause for concern when inspected, which is a testament to the quality of its construction.

Building work however is not limited to bridges alone as the stations have required a tremendous amount of time and effort spent on them in order to restore them to their original condition. While the exteriors of the buildings have remained the same, many of the interiors have had to be changed in order to cater for visitors and railway enthusiasts rather, than commuters and travellers. Rothley, the most recent station to be added to the Great Central preserved line, has had one of its waiting rooms converted into a shop and refreshment room which does a brisk trade in the summer months. At Quorn & Woodhouse the same principle has been applied and both stations are a fine example of what Great Central country stations used to look like. Loughborough Central, possibly the largest station in preservation has required an extensive amount of work to bring it up to its present splendour.

Structurally, the canopy has been painstakingly restored with rotten timbers, broken panes of glass and missing lead being replaced. This work was completed a number of years ago and now freshly painted it reflects how magnificent a structure it is.

In order to bring the staircase up to standard a major operation is being carried out to replace rotten steps leading to the booking hall and to repair the supporting timbers. Once this job is finished and painted the whole station will be an example of preservation at its best. The booking hall was one of the first parts of the station to be restored, with a complete stripping of its wooden panelling and restoration in a varnished teak. Underneath the booking hall lies one area which required structural alteration, this being the Museum which was opened in 1980 following a shortage of space in its old premises.

The permanent way maintenance, along with that of bridges and stations is only half the story as far as the operation of a railway is concerned. Some sort of control on train movements in the yard at Loughborough Central and on the main line had to be introduced and this is gradually being achieved by the Signal and Telegraph Department. One of the earliest tasks done by the group was to rebuild the signal box at Loughborough Central and establish a telegraph system between Loughborough, Quorn and Rothley. Many of the poles had to be replaced which carried the telegraph wires at Loughborough due to them being life expired, while at Quorn the need for such poles has now been obviated by the introduction of a troughing route which carries some ten pairs of telecommunications cable. Connected into the telephone system is the shed and signalbox as well as several of the rooms on the station at Loughborough. The actual signalling at Loughborough was carried out in three stages as follows:

On a dull summer Wednesday in 1980, platform 2 is seen from the road overbridge at Loughborough Central. Showing the restored canopy while, in the foreground are some coaches representing the three different periods of Great Central operation, in the form of 'Barnum' saloon, BR Mk1 SK and Gresley Coach CK No.18033. *(Author)*

Stage 1 Signalling Commissioned on 25th and 26th July 1981

This covered the north crossover, 'down' sidings points and locomotive sidings trap points and all associated signals which involved five disc ground signals, two shunt ahead signals, one miniature arm signal, two fixed stop signals and a fixed distant. Under this programme, 'Stop and await instructions' and a 'Stop proceed with caution' illuminated noticeboards were provided from the locomotive shed to the 'down' through sidings. Telephones were installed at the locomotive sidings exit at the north end of platform 1 and on Loughborough station itself.

To the south of the station a ground frame was provided with a bell and telephone for communication with Loughborough signal cabin, prior to the point being motorised at the south end of the loop through platform 2. The signalling of the south end of the station was to form Stage 2 of the signalling programme.

Stage 2 Signalling commissioned 12th May 1984

The basic aim of this part of the three stage signalling scheme was to extend the signalling control to the A6 roadbridge, thus allowing two train operation to take place safely. Train one, would be on the line between the A6 roadbridge and Rothley, while train two would be standing at Loughborough waiting for its locomotive to run round, it would then depart after train one had returned and the single line operating staff had been handed over to the driver of train two. This had the benefit of improving turn round time and making shunting in the station area easier.

With the completion of Stage 2 the ground frame was controlled by the signal cabin and a ground colour light signal was added to the frame by the point at the south end of the loop, which enabled trains to be called into platform 2. Previously the ground frame controlled point was operated by the train staff token. During the installation of equipment for Stage 2 some 13 working semaphore signals, 4 points, 2 trap points, 6 track circuits, 7 signal post telephones, 4 telephones connected to the signal box, 8 electric lever locks in the signal box, 18 clocks and 20 dial telephones were all put in. As far as wiring went, for this project 11 miles of overhead wire were put up and 1,800 metres of wire were installed in the signal box for the electric locking.

Stage 3 Signalling commissioned May Bank Holiday 1985

Stage 3 really provided the finishing touches to what has proved to be a comprehensive signalling programme which covers every eventuality as far as safety is concerned, including failure of equipment, possible human error and any other occurrence. In the final stage work centred on the provision of a point motor for the turnout south of Loughborough Central station and the commissioning of the running signals to control trains into and out of platform 2. To effect this, slight changes had to be made to the mechanical locking system in the signal box. With the completion of this task the Loughborough programme was finished, thus enabling work to take place on the signalling at Quorn & Woodhouse and Rothley.

Whilst work was being carried out on the Loughborough signalling scheme the foundations were being prepared for the new signalbox at Rothley which had been acquired from Blind Lane on the former Great Western/Great Central line in 1977. It was transported to Loughborough in pieces and over a two year period it has been reassembled opposite the 'down' platform. The signalbox is now nearing completion though no actual signal or control equipment has yet been installed.

The signal cabin at Quorn & Woodhouse however, which arrived from Market Rasen in January 1984, has also received some attention. More work has been done here though on the installation of signal equipment with the ground frame for the Quorn Loop and some signal posts already installed. The majority of signalling has been provided from the purchase of a number of redundant signals from British Railways' Toton depot. Originally it was intended that a dismantled box from Hornsey would provide the control for Quorn – this containing a lever frame which, along with one or two other furnishings from the cabin, is now displayed in the Museum.

The Museum is one asset which the railway possesses that creates a tangible link with the past glory of the old Great Central Railway and all the artefacts it left behind when it ceased to exist as a company. In the early days of the Railway's operation the Museum was a small affair comprising of just a few exhibits. By 1979 it was clear that a larger room was going to be needed to house the ever-growing number of donations, so it was decided to make use of the old information and appeal office housed at the bottom of the old lift shaft. This room provided one of the two rooms later to be designated the Museum. The second room lay behind the old office and needed to have a doorway into it and this was achieved by cutting a hole in the back wall of the old information office. The Museum was finally completed shortly before Easter 1980 and laid out with several newly acquired exhibits. The entrance hall featured a display of pictures of the building

Inside the small relics museum showing some of the cast iron and enamel signs on display. On the lower half of the wall facing the camera is a GCR colliery map, some GCR handbills, wagon plates and five name plates from steam locomotives. Underneath these is a working model of an 8B class Great Central locomotive. The wall on the right shows the Great Central in pictures through the various stages of its life while the cabinet to the left contains crockery, badges, rule books and other small artefacts.
(D. Bonas)

of the Great Central London Extension from the Newton Collection as well as a numerous examples of rail chairs and mile posts. The main room was decked out with a fine array of pictures, enamel signs, cast iron signs and relics from many different railway companies.

One of the highlights of the inner room is the area with five nameplates from locomotives attached to it, these being *The Quorn*, *Leicester City*, *Benachie*, *Sir Villiars* and *Jutland.* As the years have progressed several changes in the collection have been made as more and more items have been loaned or donated to the Museum.

The Museum has proved to be one of the Railway's greatest attractions to both visitors and members alike, and because of its popularity it now requires three volunteer members of staff to run it efficiently. The reputation alone of the Museum and

And finally, the author in Great Central uniform poses for the camera outside the Museum at Loughborough Central of which he is the Curator. ***(John Clarke)***

the trains that run on the Great Central preserved line however, cannot be responsible for attracting the necessary revenue to make the enterprise a going concern. Instead the latter must be backed up by a full scale advertising and publicity campaign, similar to the one excercised by Sir Sam Fay, General Manager of the original Great Central Railway Company. At present the advertising campaign has a series of successful posters doing the rounds as shown.

RETURN TO YESTERDAY

Give the family a return trip to yesterday on a proper steam-hauled passenger train through the delightful countryside between Loughborough and Rothley, on the preserved Great Central Railway. Enjoy a drink or a snack in the ex-Flying Scotsman buffet car (Marston's Ales), or book in advance for lunch or afternoon tea in the luxurious and nostalgic atmosphere of the first-class dining car.
After your journey, visit the museum, souvenir shop, loco shed, possibly even the signal box at Loughborough. All are run in the best traditions of railway service, completely by volunteers.
Trains run at 10.30, 11.40, 1.00, 2.30, 3.40 and 4.50 every Saturday, Sunday and Bank Holidays.

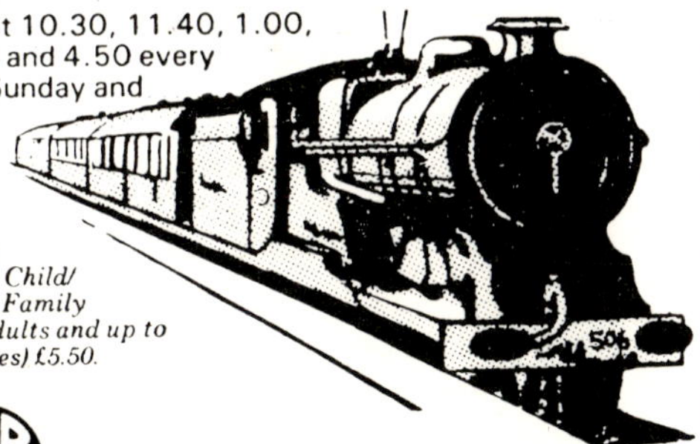

Return fares: Adults £2.00. Child/ O.A.P. £1.00. Family ticket (two adults and up to three half-fares) £5.50.

GREAT CENTRAL RAILWAY
Great Central Station, Loughborough, Leicestershire. Tel: Loughborough 230726.

CLASS 14 WEEKEND – OCTOBER 30th & 31st. 1982

The D9500 Series
A chance to travel behind the last class of diesel hydraulic locos built for British Rail — double-headed! Two of these centre-cab 650hp locomotives, D9516 and D9523, will be running on the Great Central Railway on 30th/31st October.

Trains at 10.20, 11.45, 1.30, 2.30 and 3.40. The round trip fare between Loughborough and Rothley is £1.80 for adults and 90p for children/o.a.p.'s. Family tickets covering two adults and up to 3 half-fares will be available at £4.80.

In addition, a mail train, setting down mail at speed at Quorn and Woodhouse station will be run at 12.30 on both days: this train will be steam-hauled.

Come and ride behind an unfamiliar class in an unusual setting!

GREAT CENTRAL RAILWAY
Great Central Station, Loughborough, Leicestershire. Telephone: Loughborough 30726.

DON'T MISS THE STIRLING SINGLE'S LAST STEAMING DATES

Patrick Stirling's famous G.N.R. Number 1, built in 1870 and now part of the National Collection, will be making its farewell appearances in steam at the Great Central Railway on May 8/9 and on June 12/13. These weekends will be your last opportunity to travel behind this famous locomotive in this country, possibly until the 21st century.

Trains at 10.20, 11.45, 1.05, 2.30, 3.40 and 4.50. **Accommodation will be strictly limited.** Make sure of your seat today. The round trip fare between Loughborough and Rothley, including booking fee, is £2.25 for adults and £1.25 for children. Write now to make your reservation, giving the date and first and second choice of train time for your trip.

Payment must accompany all bookings but Access or Visa cardholders may book by phone, quoting their card number.

GREAT CENTRAL RAILWAY
Great Central Station, Loughborough, Leicestershire. Telephone: Loughborough 30726.

Please send S.A.E. for further details to:
The General Manager,
Great Central Railway,
Loughborough, Leicestershire LE11 1SS.